Springer Tracts on Transportation and Traffic

Volume 11

Series editor

Roger P. Roess, New York University Polytechnic School of Engineering,
New York, USA
e-mail: rpr246@nyu.edu

About this Series

The book series "Springer Tracts on Transportation and Traffic" (STTT) publishes current and historical insights and new developments in the fields of Transportation and Traffic research. The intent is to cover all the technical contents, applications, and multidisciplinary aspects of Transportation and Traffic, as well as the methodologies behind them. The objective of the book series is to publish monographs, handbooks, selected contributions from specialized conferences and workshops, and textbooks, rapidly and informally but with a high quality. The STTT book series is intended to cover both the state-of-the-art and recent developments, hence leading to deeper insight and understanding in Transportation and Traffic Engineering. The series provides valuable references for researchers, engineering practitioners, graduate students and communicates new findings to a large interdisciplinary audience.

More information about this series at http://www.springer.com/series/11059

Francesc Soriguera Martí

Highway Travel Time Estimation With Data Fusion

 Springer

Francesc Soriguera Martí
Barcelona Tech
Technical University of Catalonia
Barcelona
Spain

ISSN 2194-8119   ISSN 2194-8127 (electronic)
Springer Tracts on Transportation and Traffic
ISBN 978-3-662-48856-0   ISBN 978-3-662-48858-4 (eBook)
DOI 10.1007/978-3-662-48858-4

Library of Congress Control Number: 2015954978

Springer Heidelberg New York Dordrecht London
© Springer-Verlag Berlin Heidelberg 2016
This work is subject to copyright. All rights are reserved by the Publisher, whether the whole or part of the material is concerned, specifically the rights of translation, reprinting, reuse of illustrations, recitation, broadcasting, reproduction on microfilms or in any other physical way, and transmission or information storage and retrieval, electronic adaptation, computer software, or by similar or dissimilar methodology now known or hereafter developed.
The use of general descriptive names, registered names, trademarks, service marks, etc. in this publication does not imply, even in the absence of a specific statement, that such names are exempt from the relevant protective laws and regulations and therefore free for general use.
The publisher, the authors and the editors are safe to assume that the advice and information in this book are believed to be true and accurate at the date of publication. Neither the publisher nor the authors or the editors give a warranty, express or implied, with respect to the material contained herein or for any errors or omissions that may have been made.

Printed on acid-free paper

Springer-Verlag GmbH Berlin Heidelberg is part of Springer Science+Business Media
(www.springer.com)

*As a general rule the most successful man
in life is the man who has the best information*

Benjamin Disraeli

Preface

Overview, Objectives, and Main Contributions

Overview

In the present context of restraint in the construction of new infrastructures due to territorial, environmental, and economical restrictions, and given the big bang of mobility in the last decades in all sectors, the transportation system management acquires a fundamental role as an optimizer of the available resources. This requires the application of new policies addressed to achieve two main objectives: sustainability and competitiveness.

Mobility management is the big issue. The information technologies and communications are the tools. It is therefore needed a management system that links these objectives with these available tools. This management system should be based on quantitative traffic information in real time. Travel time and its reliability stand out as key factors in traffic management systems, as they are the best indicators of the level of service in a road link and perhaps the most important variable for measuring congestion. In addition, travel time is the best and most appreciated traffic information for road users as it plays a fundamental role in the traveler planning process (travel or not travel, best time to travel, route and mode choice…). At the same time, highway travel time measurement and quantitative forecasting in congested conditions pose a striking methodological challenge.

The most elementary method to measure a highway travel time is by identifying the vehicle at the entrance and exit of the target section and computing the elapsed time between identifications (i.e., direct measurement). However, the necessary automatic vehicle identification (AVI) is not a trivial task. As it will be described in this book, it needs somehow advanced technology. All of these technologies require the extensive installation of new hardware in the vast highway network, which cannot be achieved overnight, and possibly, it is not profitable in the whole network.

To date, these systems have only seen limited demonstration, and the extensive deployment is not expected beyond the hot spots of the network.

Loop detectors still represent the main source of traffic data in all highways worldwide. And it is expected to remain like this in the medium term (May et al. 2004). Note that loop detectors are not adequate for link measurements (e.g., travel times) but are capable of a comprehensive measurement of the punctual data (e.g., traffic counts: the original objective), which is not less valuable and for which the AVI technologies are not so well suited. Given the predominance and preexistence of this surveillance context, lots of traffic agencies worldwide have decided to develop travel time information systems based on a simple and intuitive methodology: They estimate indirectly link travel times from the spatial generalization of the loop punctual measurement of average speed.

As it will be detailed in the next chapters, in both the direct measurement and the indirect estimation of highway travel times difficulties arise. In addition, different measurement processes lead to conceptually different results (see Chap. 2 devoted to travel time definitions), an issue which is frequently overlooked. Note that a directly measured travel time is a trajectory-based measurement in space-time where the vehicle needs to have finished his trip in order to obtain the measurement. In contrast, indirect estimations usually are instantaneously obtained and do not respond to the trajectory of any particular vehicle. If that was not enough, the main objective of a real-time highway travel time information system should be to provide the driver with the information of the travel time his trip will undertake once entering the highway. This means that a real-time information system needs in reality future information, where the horizon of the forecast is equal to the trip travel time.

This situation, with multiple surveillance equipments, inhomogeneous data with different variables being measured, different travel time estimation algorithms with different accuracies, different spatial coverage, and different temporal implications, is the ideal environment for data fusion schemes, where the objective is to use jointly the information provided by different sources in order to infer a more accurate and more robust estimation of the target variable (i.e., the travel time).

Objectives

The main objective of the present monograph is to present a methodology capable of providing the driver entering a highway with accurate information of his expected travel time. Note that this information involves two components: the measurement of the current travel times and the estimation of the evolution of the traffic conditions during the time taken for the trip. An additional requirement is that the travel time estimation must be obtained by making the best usage possible of the available and multiple highway data sources, neither increasing the highway density of surveillance nor changing the typology of the measurement equipment.

Therefore, the objective is to add value to the traffic data which are currently being measured.

Two main research directions appear when facing the issue of real-time monitoring of the traffic evolution. The first and most intuitive way of quantitatively knowing what is happening in a highway stretch is by measuring. In practice, this is not an easy task. The measurement equipments are limited. They may not be able to measure the most important variables. The amount of measurements may not be representative of the average traffic stream. Their spatial coverage may be limited. The necessary temporal aggregations to reduce the amount of data being transmitted may bias the measurement and add some delay to the information. The existence of outliers adds additional complexity. And finally, a non-negligible amount of measurement units may usually be out of order. The alternative consists in modeling. Highway traffic consists in the interaction of a huge number of drivers: human beings with different ages, different races, different religious creeds, different gender, different political preferences, different way of life options and different psychological stabilities (Vanderbilt 2008). Nevertheless (and perhaps surprisingly), on the average, they all behave similarly when they face similar conditions. This means that, given the characteristics of the drivers' population, the characteristics of the highway environment they are facing, and the mobility demand (i.e., the number and characteristics of the trips), it should be possible to know how these vehicles are going to interact and therefore to obtain all the resulting variables of their trips (e.g. their travel times). The forecasting capabilities of this approach are appealing. Again, this is not an easy task. It is difficult to know (which means measure) the characteristics of the drivers' population. And it is difficult to accurately model the behavioral laws that steer the relation between the infrastructure and drivers and also between drivers among themselves. In fact, some of these behavioral laws are still unknown (Daganzo 2002). It is even difficult to know the mobility demand. In conclusion, there are a lot of unknowns.

The dilemma is then to select one option. Lately, modeling has experienced an enormous popularity increase. It seems that modeling is now easier than has ever been before. And the results are better (at least faster—in real time—and more visual). This is due to the quick development of computers and the enhancement of its visual capabilities, which has given rise to traffic simulators (in particular microsimulators, where the performance of each vehicle can be seen in 3D). Although these enhancements brought up by the digital era, the baseline difficulties remain the same. One must realize that most of these difficulties are solved by case-specific overcalibrations, which blur the forecasting capabilities of the model. Fortunately, researchers all over the world are working hard in overcoming these problems, and in the future, this may come to a happy end. In contrast, traffic monitoring seemed to have fallen out of favor some years ago. This was mainly due to the huge costs of enhancing the traffic surveillance systems given the enormous inertia of the vast highway network. The surveillance equipment rapidly became outdated, with high maintenance cost and a high rate of malfunctioning. This entire situation discouraged practitioners and researchers from devoting their interests in traffic monitoring. This is now changing with the appearance of high-tech low-cost traffic detectors. Technological reliability will surely be enhanced, and acquisition and maintenance costs will be cut down. However, the conceptual difficulties in the measurements will also remain the same. In addition, all the surveillance system will not be replaced overnight, and different equipment will need to coexist.

Both research directions, measuring and modeling, are appealing. Both have a huge potential. In spite of the modeling "attacks" and the temporary monitoring decline, measuring will not be substituted by modeling. Traffic monitoring will be always necessary. Maybe the axiom should be "measure all you can; model the rest." This is what this book is devoted to: providing methodologies to accurately measure highway travel times. In general, methodologies are not technologically captive, as it is more related to the concept of the measurement than to the technological equipment used.

In particular, the example of application of the proposed methodologies in the book is presented in the specific case of a closed toll highway, where the direct travel time measurement is given by the information contained in the toll tickets (real of virtual by means of electronic toll collection systems), which record the exact time and location where each vehicle enters and leaves the highway. The indirect estimation is obtained from the flow, speed, and occupancy measurements of inductive loop detectors. Although this specific environment of application, the proposed methodologies will be easily generalized to other context where both a direct measurement and an indirect estimation are available.

When facing the travel estimation problem in this closed toll highway environment, with these commonly available sources of data, three main questions arise:

- How travel time can be measured from toll ticket data?
- How travel time can be measured from loop detector data?
- Given these two travel time estimations from different data sources with their intrinsic characteristics, can we combine them to obtain better information?

These questions are going to be answered in the present book.

Structure and Main Contributions

After this overview of the monograph, which gives an introduction to its contents and provides linking arguments between its different parts, the rest of the book is structured as a compendium of seven self-contained chapters. Each one deals with some part of the global research question. This may facilitate the partial reading and diffusion of its contents.

Chapter 1, entitled *Highway Travel Time Information Systems: A Review*, has the objective of providing a global view of the issues treated in the book and of the results obtained. Some baseline concepts are also introduced. Specifically, the issues addressed include the analysis of the importance of travel time information in mobility management, the qualitative description of direct and indirect methods for travel time estimation, the definition of data fusion concepts and their relationship with travel time forecasting, the effects of travel time information dissemination strategies on drivers, and a discussion on some issues regarding the value of travel time information as a traveler-oriented reliability measure. Finally, some overall conclusions and issues for further research are outlined.

Chapter 2 is an instrumental part of the book, where travel time definitions are analytically presented. Also, a trajectory reconstruction algorithm necessary in order to navigate between different travel time definitions is proposed. The concepts presented in this chapter are aimed to create a conceptual framework useful in comparing travel times obtained from different methodologies. This should be considered as baseline knowledge when going through the whole book.

Chapter 3 is devoted to indirect travel time estimation from loop detector data. Specifically, it addresses the main drawback in obtaining travel times from punctually measured average speed: the spatial generalization of the measurements. Several methods are proposed in the literature ranging from the simplest constant interpolation to mathematically complex truncated quadratic ones. The research tendency seems lately to follow the direction of continuously increasing the mathematical complexity of the methods overlooking traffic dynamics. This issue is addressed in this Chap. 3, entitled *Accuracy of Travel Time Estimation Methods Based on Punctual Speed Interpolations*. This chapter claims that all speed interpolation methods that do not consider traffic dynamics and queue evolution do not contribute to more accurate travel time estimations. Lacking a better approach, the simplest midpoint interpolation is recommended.

While Chap. 3 highlights the main problems of travel time estimation methods based on punctual speed measurements and proves that the inaccuracies resulting from limited measurement spots are unavoidable without the consideration of traffic dynamics, Chap. 4 claims that this improved accuracy may not be necessary for real-time information systems and in some cases may even be detrimental. This contradicts the common perception that freeway travel time information systems, whose objective is to provide real-time information, must be supported by very accurate travel time measurements. This perception leads traffic agencies to be more prone to using fancy new technologies for directly measuring travel times than to make the most of already installed loop detectors. Chapter 4, entitled *Design of Spot Speed Methods for Real-Time Provision of Traffic Information*, shows how this can be a myopic approach, as more accurate travel time measurement may lead to worse performance of a real-time information system. In addition to this claim, this chapter provides useful guidelines for practitioners for setting the main parameters of the system, and also some practice ready enhancements to commonly used spot speed travel time estimation methods. In the authors' opinion, this is a significant contribution and especially relevant and interesting for the practitioner community. It can also provide a global framework that may help researchers to not forget the final objective.

Chapter 5 deals with direct travel time measurement. Entitled *Highway Travel Time Measurement from Toll Ticket Data*, it provides a method capable of obtaining main trunk average travel times (e.g., in between junctions) from specific origin–destination individual vehicle travel times, which include the "entrance time" and the "exit time" (i.e., the time required to travel along the on-/off-ramp and to pay the fee at the toll plaza. This method allows reducing the intrinsic delay in the information of directly measured travel times, which is essential for a real-time application of the information system.

Having obtained travel time estimations from the different available data sources (i.e., direct and indirect), Chap. 6 entitled *Short-Term Prediction of Highway Travel Time Using Multiple Data Sources* proposes a data fusion scheme, partially based on the probabilistic Bayes' theory, whose objective is to use the potentials of each source of data to overcome the limitations of the others in order to obtain a more accurate and robust travel time estimation. In addition, the proposed method uses the different temporal alignments of travel time estimations to infer a tendency and improve the forecasting capabilities of real-time measurements. The source estimation methods used in this last chapter are the ones presented in the previous chapters.

Chapter 7, entitled *Value of Highway Information Systems*, is the last chapter of the monograph. Its main objective is to assess the real benefits of information systems, widely implemented worldwide. Contributions of the chapter can be grouped into two aspects. First is the results, quantifying the value of travel time information in different scenarios. These include one or two available routes, peak or off-peak traffic, different types of trips, and massive or limited dissemination of strategies. The richness of this set of scenarios overcomes the limitations of other research efforts of the same nature. Second is the methodology, proposing a departure time selection model based on a simplification of the expected utility theory, with some restrictions to account for already planned decisions and habits, and a cost model accounting for the unreliability of the trip. This is based on Small's classic model with some modifications to include stress and the possibility of rescheduling activities as a result of information. The empirical data presented for the application of the method (measured in a Spanish highway) may also be a valuable contribution. All these arguments make this chapter interesting for both researchers and practitioners.

Summarizing, the objective of the book is to provide solutions to a global engineering problem. Each one of the chapters provides answer to a partial question, which follows from the original research problem. The unity of the topic treated is therefore granted.

In addition, in any engineering monograph, the practical application of the proposed methodology to a pilot test site is desired. In this particular case, data from a privileged site were available. The AP-7 highway runs along the whole Spanish Mediterranean coast, from Algeciras to the French border at La Jonquera. On the northeastern stretch of the highway, from La Roca, near Barcelona, to La Jonquera, a closed tolling scheme is in use. Toll ticket data were available to the author. Moreover, in some sections of the highway near Barcelona (in particular from La Roca to Maçanet—see Fig. 1), additional monitoring by means of loop detector data was installed every 5 km approximately. Only a requirement is missing for this stretch being a perfect test site: a congestion episode. Unfortunately, for the highway users (but fortunately for the development of the present book), every Sunday

Fig. 1 Test site location. *Source* Google Maps

(among other days) of the summer season (particularly long in the Mediterranean climate), congestion grows in the southbound direction of the highway, due to the high traffic demand toward Barcelona of drivers which have spend a day or the weekend on the coast. Hopefully, the contents in this book may help to alleviate this congestion, or at least it will provide information to diminish the drivers' suffering. This was the privileged test site used in all the research presented here.

References

Daganzo, C. F. (2002). A behavioral theory of multi-lane traffic flow. *Transportation Research Part B, 36*(2), 131–158.

May, A., Coifman, B., Cayford, R., & Merrit, G. (2004). *Automatic diagnostics of loop detectors and the data collection system in the Berkeley Highway Lab*. California PATH Research Report, UCB-ITS-PRR-2004-13.

Vanderbilt, T. (2008). *Traffic: Why we drive the way we do (and what it says about us)*. New York, Toronto: Alfred. A. Knopf publisher.

Contents

1 Highway Travel Time Information Systems: A Review 1
   1.1   Travel Time and Mobility Management 1
   1.2   Travel Time Measurement............................. 3
         1.2.1   Direct Travel Time Measurement................. 4
         1.2.2   Indirect Travel Time Estimation................. 12
   1.3   Data Fusion and Travel Time Forecasting................. 16
         1.3.1   Data Fusion Schemes......................... 19
         1.3.2   Bayesian Data Fusion......................... 21
         1.3.3   Main Benefits and Drawbacks of Data Fusion
                 Schemes....................................... 23
   1.4   Travel Time Information Dissemination 25
         1.4.1   Information Before Departure 25
         1.4.2   Information En Route........................... 27
         1.4.3   Comparison of Different Traffic Information
                 Dissemination Technologies 28
   1.5   Value of Travel Time Information 29
         1.5.1   Travel Time: Variability, Reliability and Value
                 of Information................................ 30
         1.5.2   Travel Time Unreliability Expected Behavior
                 in Multilane Freeways 33
         1.5.3   Sources of Travel Time Unreliability:
                 What Could Be Done?.......................... 34
         1.5.4   Travel Time Information System: Levels
                 of Application in a Road Network 35
   1.6   Conclusions and Further Research 36
   References... 39

2 Travel Time Definitions 43
   2.1   Introduction.. 43
   2.2   Link Travel Time Definitions 44

		2.2.1	When Travel Time Definition Makes a Difference	46
		2.2.2	Which Information Is Actually Desired from Real Time Systems?	48
	2.3	Corridor Travel Time		48
	2.4	Trajectory Reconstruction Process		49
	References			52

3 Accuracy of Travel Time Estimation Methods Based on Punctual Speed Interpolations 53
   3.1   Introduction and Background. 54
   3.2   Methods of Link Travel Time Estimation from Point Speed Measurements 58
         3.2.1   Constant Interpolation Between Detectors 59
         3.2.2   Piecewise Constant Interpolation Between Detectors.... 61
         3.2.3   Linear Interpolation Between Detectors. 63
         3.2.4   Quadratic Interpolation Between Detectors 64
         3.2.5   Criticism to the Presented Methods 67
   3.3   The Data. 68
   3.4   Evaluation of Proposed Speed Spatial Interpolation Methods.... 72
   3.5   Conclusions and Further Research 79
   References 82

4 Design of Spot Speed Methods for Real-Time Provision of Traffic Information 85
   4.1   Introduction. 85
   4.2   Design of Spot Speed Methods for Real-Time Provision of Traffic Information 88
         4.2.1   Immediacy Requirement in Rapidly Evolving Traffic Conditions 90
         4.2.2   Midpoint Algorithm Forecasting Capabilities 92
         4.2.3   Applications of Other Constant Speed Methods: The Range Method. 94
   4.3   Spot Speed Methods Improvement in Stationary Conditions 96
         4.3.1   Intelligent Smoothing Process 96
         4.3.2   Granularity in the Disseminated Information 98
         4.3.3   Stationary Conditions: Opportunities for Data Fusion Schemes. 99
   4.4   Test Site and Empirical Data. 100
   4.5   Empirical Results. 102
   4.6   Conclusions 105
   References 106

5 Highway Travel Time Measurement from Toll Ticket Data 109
   5.1   Introduction. 109
   5.2   Objective of the Proposed Algorithm 112

	5.3	Estimation of Single Section Travel Times: The Simple Algorithm's Underlying Concept	115
		5.3.1 Basic Algorithm	116
		5.3.2 Extended Algorithm	119
	5.4	Modifications for the Real Time and Off-Line Implementations of the Algorithm	122
		5.4.1 Real Time Implementation	122
		5.4.2 Off-Line Implementation	125
	5.5	Application to the AP-7 Highway in Spain	127
		5.5.1 Selection of the Sampling Duration "Δt"	129
		5.5.2 Accuracy of the Algorithm	133
		5.5.3 Value as a Real Time Information System	136
		5.5.4 Exit Time Information	137
	5.6	Conclusions and Further Research	138
	References	153
6	**Short-Term Prediction of Highway Travel Time Using Multiple Data Sources**	157
	6.1	Introduction	158
	6.2	Travel Time Definitions Revisited	160
	6.3	Naïve Travel Time Estimation Algorithms	162
		6.3.1 Spot Speed Algorithm for Travel Time Estimation	163
		6.3.2 Cumulative Flow Balance Algorithm for Travel Time Estimation	165
		6.3.3 Travel Time Estimation from Toll Ticket Data	168
	6.4	Data Fusion Methodology	169
		6.4.1 First Level Data Fusion	170
		6.4.2 Second Level Data Fusion	172
	6.5	Application to the Ap-7 Highway in Spain	175
		6.5.1 First Level Fusion Results	176
		6.5.2 Second Level Fusion Results	179
	6.6	Conclusions and Further Research	182
	References	182
7	**Value of Highway Information Systems**	185
	7.1	Introduction and Background	186
	7.2	Modeling Departure Time and Route Choice	191
		7.2.1 Including Perception Errors	192
		7.2.2 Effects of Information	193
	7.3	Modeling Travel Time and Unreliability Costs	197
		7.3.1 Value of a Travel Time Information System	199
		7.3.2 Morning Commute: Equivalence with the Expected Utility Theory	200

7.4	Model Limitations and Some Solutions		201
	7.4.1	Limited Dissemination of the Information	201
	7.4.2	One Route, Massive Dissemination of Information	201
	7.4.3	Two Routes, Massive Dissemination of Information	202
7.5	Numerical Examples		204
	7.5.1	Model Results	206
7.6	Summary and Conclusions		209
References			211

Chapter 1
Highway Travel Time Information Systems: A Review

Abstract This chapter starts analyzing the importance of travel time information in mobility management. After that, it presents a qualitative description of direct and indirect methods for travel time estimation and defines data fusion concepts and their relationship with travel time forecasting. The effects of travel time information dissemination strategies on drivers and a discussion on some issues regarding the value of travel time information as a traveler oriented reliability measure are also presented. Finally, some overall conclusions and issues for further research are outlined.

1.1 Travel Time and Mobility Management

Mobility is synonymous of economic activity and dynamism. It has been vastly proved the relationship between mobility demand and the wealth of a particular region (Robusté et al. 2003). The mobility increase implies greater competiveness and, if properly planned, territorial cohesion. However, for the mobility to provide these benefits, a good transportation network and a better management of the transportation system is necessary. An infrastructural deficit or the absence of an active management may entail the increase of mobility being counterproductive, transforming the potential benefits to additional costs. These over costs are mainly due to the congestion phenomena.

Congestion is linked to success. It appears when the interaction between transportation demand and the transportation supply of the system (in terms of infrastructure and organization) generates increasing unitary costs to overcome the same unitary length. Taking into account that the infrastructural supply can hardly go ahead of the demand, given the limitation of resources, congestion has to be considered as an inevitable phenomenon which indicates success and acts as a demand regulator. In spite of this, congestion must be managed and must be sustained as punctual and moderate episodes: it is necessary to maintain a "suitable" level of congestion. The first step is then to know and quantify the level of congestion and to try to limit its damaging variability. This means that for the same trip on two similar days, the travel time should be similar, not the double. This concept is known as travel time reliability.

In most of the metropolitan areas worldwide, the existing levels of congestion are far above from these suitable thresholds. In addition, and despite the actual context of economical recession which has alleviated the growing trends, metropolitan congestion in developed areas is still slightly increasing (Federal Highway Administration 2010). This is translated into huge social costs.

When facing this situation of growing congestion in metropolitan areas, two main approaches exist to alleviate the problem: to increase the amount of infrastructures, or to improve the management of the existing ones. Usually, the construction of new metropolitan freeways is only a temporary solution, as involves more induced traffic (and more congestion), plus an increase in the territorial occupancy, already severely harmed. It is not possible to maintain a continuous increase in the infrastructural supply, due to the funding limitations, but mainly due to lack of sustainability of this approach, given the difficulties in obtaining a respectful territorial integration. The capacity of territory to absorb new infrastructures is finite. These assertions do not mean that the current infrastructures must remain still. Some regions surely need more construction. And some others may need the reconstruction of the infrastructures, in order to adapt them to more sustainable urban transportation modes. Transportation infrastructures must be capable of a continuous adaptation to the needs of the society.

The alternative is the improvement of traffic management. This may imply actions to modulate the demand (e.g. increase of vehicles' occupancy, smooth peak hour demands, derive demand to other transportation modes) usually by means of taxation or restriction. And, in addition, a better management of the supply (e.g. improving the lane usage, avoiding traffic instabilities by imposing variable speed limits, avoiding the capacity drop by imposing ramp metering, ...). It is in this context where a common baseline requirement appears: traffic information. Traffic information is needed by traffic managers in order to set their operational policies. It is also needed by drivers in order to take their own decisions (Fig. 1.1).

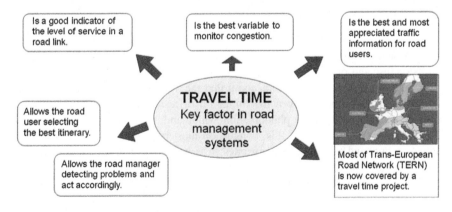

Fig. 1.1 Travel time importance in the mobility management context

Travel time information appears as the key element. Travel time is the fundamental variable to provide traffic information, because is the best indicator of the level of service in a road stretch and it is completely understandable by all users. In fact, several surveys have shown that travel time is the worthiest information from the user perspective (Palen 1997), as it allows him to decide in advance when is the best time to start a trip and the best routing option, or to modify this initial planning once on route. Travel time information is not only useful to the driver, but also to the road system operator as it is a basic knowledge to assess the operational management and planning of the network. Travel time forecasting allows the operator/manager to beat the incidents and the operational problems in the system, while the real time information allows monitoring the evolution of these incidents. The network manager must not only provide travel time information to drivers, but also look after its variation and achieve high reliability of the infrastructure. In this context, travel time measurement and forecasting must be a priority objective for road network managers and operators.

1.2 Travel Time Measurement

The need for traffic data appeared as soon as motor vehicles became popular and the road network started to develop, almost 100 years ago. Originally, the objective was limited to measuring traffic volumes in order to know the usage of the network for planning purposes. Soon thereafter average speeds became necessary to estimate the level of service of the different links and enhance the planning process (Highway Research Board 1950). More recently, with the development of the information and communication technologies, real time traffic data play the main role in the so-called "Highway Advanced Traffic Management and Information Systems" (ATMS/ATIS) (Palen 1997). At every step of this evolution, the requirements in terms of equipment, communications, processors … in short, cost, to fulfill the traffic data needs has increased enormously. In addition the vast extension of the highway network adds a huge inertia in relation to the surveillance system already installed, so that it cannot make the most of the continuous technological and economical improvement of equipment. These factors lead to very heterogeneous levels of surveillance within a highway network. On the one hand the hot spots of the network (e.g. metropolitan freeways) may be densely monitored, with various types of equipment, including high-tech. On the other hand, some parts of the network may remain with some isolated outdated detectors. All surveillance levels could be found in between. Obviously, metropolitan freeways concentrate most of the traffic of the highway system (and therefore most of the operational problems) and the intensive monitoring is completely justified (OECD/JTRC 2010).

Highway travel time estimation reflects this reality. Although being the most primitive variable to be measured in any trip (Berechman 2003), the systematic measurement of travel times in the network is quite recent. Travel time information is considered to be the key factor in the ATMS/ATIS, as it is widely accepted that the real time knowledge of highway travel times is the most informative traffic variable for both, drivers (it is easily understood, allows supporting trip decisions like changing the time of departure, switching routes or transportation modes...) and traffic agencies (it is a clear indicator of the level of service provided). This belated blossoming of highway travel time information systems may be result of the difficulties in the systematic direct measurement of travel times.

Basically there are two methodologies to measure travel time in a road link: the direct measurement and the indirect estimation. The direct travel time measurement is based on measuring the time interval that a particular vehicle takes to travel from one point to another. The alternative is the indirect travel time estimation from traffic flow characteristics (density, flow and speed), obtained for example from inductive loop detectors. To obtain travel time estimations from these last measurements some type of algorithm must be applied. Indirect measurement is especially interesting when direct measurement is extremely difficult or costly, or when all the monitoring equipment for indirect estimation is already available, while for direct measurement it is not.

Direct travel time measurement has only been carried out under regional travel time research projects, mainly in USA and Western Europe in a limited selection of corridors. With few exceptions, the availability of valid travel time databases is very limited. This lack of ground truth data has been a recurrent problem for practitioners in developing and validating their road management schemes.

Taking into account that travel time is the preferred information for all the stakeholders (managers and users), filling the gap of ground truth data should be a main objective. Although being aware of the problem, efforts in direct measuring travel times have been very rare until recent times. This results from the traditional point of view in relation to the traffic data, considered only useful for pavement maintenance and planning objectives. As a consequence, all operational usage of traffic data relies on data gathered with very different objectives and with unsuitable accuracy requirements.

1.2.1 Direct Travel Time Measurement

Travel time can be directly measured from the vehicles travelling on the highway. One of the main properties of directly measured travel times is their spatial implication: the measurements include all the effects suffered by the vehicle while traveling along space. This is a main difference from indirect estimations, which, as will be seen next, are generally based on punctual measurements and spatial extrapolations.

In the direct measurement, the travel time is an individual property of each vehicle. This means that in order to obtain a representative average travel time for a particular section, a significant number of vehicles must be measured within the traffic stream. This usually represents a drawback in this type of measurements.

1.2 Travel Time Measurement

There are two main procedures to obtain travel time as a direct measure: identifying the vehicle in at least two control points or following the vehicle along with its trajectory. Both are analyzed in the following sections.

1.2.1.1 Vehicle Identification at Control Points

In the identification based techniques, the vehicle is identified at the entrance and at the exit of the stretch, and its passing time is stored. By pairing both registers travel time is directly obtained. Obviously, clock synchronization at control points is a major issue in order to warrant the accuracy of measurements. A collateral benefit of this travel time measurement method is that the individual identification of vehicles allows for constructing origin—destination matrices a key input for simulation models which usually is very difficult to obtain.

Travel time measurement in this case responds to a particular trip of a vehicle. Therefore, it has to be finished for being measured. This time alignment implication, analyzed in more detail in Chap. 2, involves some delay in the real time application of travel time information systems. This type of measurement is generally named Arrival Based Travel Time (ATT).

Another drawback of reidentification methods is that travel times are spatially captive of the control point locations. Travel times can only be obtained between two control points. No partial measurements can be obtained. It is evident that the number and location of control points plays an important role. For high control point densities, all the drawbacks are less dramatic: information delay is slight and sections are so short that no partial information is desired. However the installation and maintenance costs increase. A trade-off must be reached. In this optimization process, not only the number of control points matter, also its location in relation to the mobility patterns, has implications. Some particular locations imply more added value to the system, while others may be irrelevant (Sherali et al. 2006). As an order of magnitude control points are located approximately every 2 km in metropolitan freeways with a high density of junctions while in interurban freeways with fewer junctions they are located up to 8 km apart (Turner et al. 1998).

In practice, more difficulties arise. For instance a common difficulty encountered when directly measuring travel times between control points is the elimination of "outliers". Only travel times directly related to the traffic conditions should be considered. Other factors, not related to traffic, introduce "false" delay to some vehicles (e.g. stopping for refueling or to have a break, or in the opposite direction motorbikes dodging congestion). If the amount of measurements is high, it is not difficult to identify these outliers using standard statistical algorithms. However, if the identification rate is low, and given the high variance on section travel times introduced in case of congestion, it is particularly difficult to discriminate, for example an episode of growing congestion from a vehicle which has stopped. This issue is analyzed in Chap. 5.

The number of identifications is crucial to obtain a representative sample, and depends basically on the identification technology. Nowadays, all systems with the objective of a systematic application must rely on the AVI (Automated Vehicle Identification) systems. Manual identification should only be considered in small specific analysis in order to avoid the implementation costs of an automatic system. Some common AVI technologies include the license plate video recognition (Buisson 2006; NYSI&IS 1970) see Fig. 1.2, the reidentification of vehicle signatures from video cameras (Huang and Russell 1997; MacCarley 2001), the identification of toll tags in the case of turnpikes (traditional toll tickets in case of closed turnpikes—see Chap. 5—or equipped with an electronic toll collection system—ETC—(Nishiuchi et al. 2006) see Fig. 1.4 or the innovative Bluetooth signature identification of on-board devices (Barceló et al. 2010).

Despite of technological malfunctioning, all vehicles could be identified in case of the license plate reading or closed toll highways scenarios. Only some of them in case of using the rest of the technologies described (depending on its penetration rate). The amount of travel time measurements depends on the technology but also on the configuration of the control points. A control point in the main highway trunk can be comprehensive when it tries to identify all vehicles crossing the section, or partial where for example only some lanes are monitored. Also note that the amount of measurements is not directly the amount of identifications. The amount of pairings between control points is the relevant parameter. In case of an exhaustive control point, the number of pairings should be almost the same as the number of identifications, despite the technology used, and if there is no on/off ramp in between. In case there is one junction, the differences respond to the originated or finished trips in the junction, which provides the data for the origin—destination matrix construction. In case of partial control points, the same cannot be asserted. The number of pairings could be significantly lower than the identifications due to the amount of "leaks" in the system. In addition, in case of an in between junction, nothing can be said about the amount of input/output vehicles. The limitations of partial control points are therefore evident. If the control points are located many kilometers apart with a considerable number of in between junctions, the origin—destination properties of the method are lost, and the amount of pairings, even in the case of exhaustive control points, will depend on the number of junctions and the number of vehicles which travel the whole itinerary. This may be a very small part of the identifications,

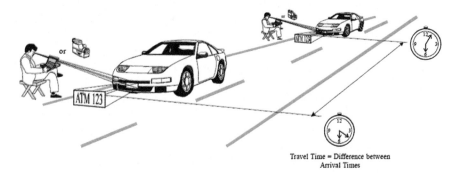

Travel Time = Difference between Arrival Times

Fig. 1.2 Travel time estimation from license plate recognition. *Source* Turner et al. (1998)

1.2 Travel Time Measurement

questioning the ability of the method for providing a continuous and significant average of travel times. As a general rule if long trips are predominant a smaller number of control points may suffice.

Finally, take into account that in some of these identification methods (e.g. license plates) it is possible to link the information with a particular person. This may imply additional legal difficulties in relation to privacy issues.

Despite the technological feasibility of automated vehicle identification, highway traffic monitoring is, and will be for the next years, based on inductive loop detectors. This results from a reminiscence of the past, where the technological options were by far more limited, and the objectives to be fulfilled by the obtained traffic data more elementary. In addition, the huge inertia implied by the vast extension of the highway network, prevents from an extensive and fast technological update. Therefore, if a highway travel time information system aims to be generally implemented in the next years, it has to be based on loop detector data.

Considering this situation, researchers have attempted to improve the travel time measurement capabilities of loop detectors by trying to reidentify vehicles at the detector spot. This allows for the direct travel time measurement. The reidentification by means of the vehicles' electromagnetic signature (Abdulhai and Tabib 2003; Kuhne and Immes 1993; Kwon 2006) see Fig. 1.3, needs retrofitting loop detector hardware. An alternative is using the vehicles' distinctive length (Coifman and Cassidy 2002; Coifman and Ergueta 2003; Coifman and Krishnamurthya 2007). However, only rare vehicles are being reidentified using these methods when lane changing and in/out flows at ramps between detectors are considered. This may add some bias to the results in free flowing situations due to the probably different travelling speed of these special vehicles, but it may not imply a serious flaw in congested conditions where FIFO traffic (i.e. First in—First out) prevails. Other approaches (Lucas et al. 2004; Dailey 1993; Petty et al. 1998), try to reidentify the platoon structure of a traffic stream, which is lost in congested periods, when travel time information is more valuable. Despite these limitations, the use of inductive loop detectors as an AVI equipment stands out as an active research field.

Direct travel time measurement in toll highways

As it is one of the main topics treated in this monograph, a little more attention will be paid at the particular context of using the equipment originally designed for collecting the toll at turnpikes for the direct travel time measurement. Travel time measurement is obtained by means of vehicle reidentification, like the rest of technologies belonging to the present category.

In closed toll highways, where the toll paid by each vehicle depends on its particular origin and destination and on the application of a kilometric fee, the vehicle reidentification is straightforward. Note that in order to compute the fee, the vehicle must be reidentified. This is achieved by means of a toll ticket, real (i.e. a piece of paper) or virtual (i.e. a register in an electronic tag) where the precise time and location the vehicle enters and exits the highway is stored. The real time exploitation of these data, which is easy as the ticket travels with the vehicle like a baton, provides the desired travel time measurements and the origin—destination

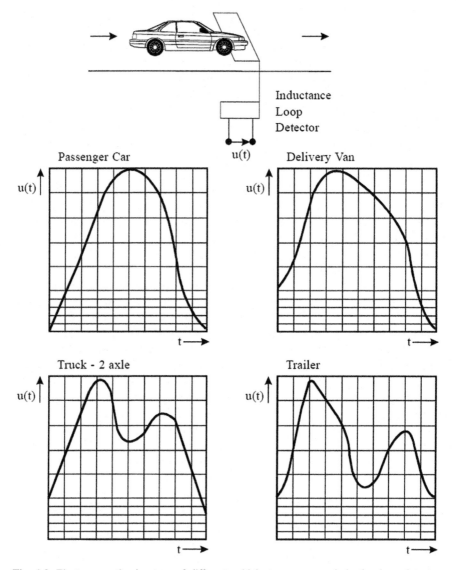

Fig. 1.3 Electromagnetic signature of different vehicle types over an inductive loop detector. *Source* Turner et al. (1998)

matrix for all vehicles. It is a valuable and exhaustive source of information usually not fully exploited.

In addition to the general problems stated before for all methods based on vehicle reidentification, this method suffers from a specific problem. The control points (i.e. the toll booths where the toll tickets are processed) are not located on the main highway trunk, but at the very far end of the on/off ramps, at junctions.

1.2 Travel Time Measurement

This means that the measured travel time from an origin to a destination includes the main trunk travel time, but also the time required to travel along the on ramp, the time required to travel along the off ramp, plus the time required to pay the toll. In short trips this additional time is not negligible. Furthermore, if one constructs itinerary travel times by adding up different single section travel times, it will add up as many entrance and exit times as sections contains the itinerary. This process may result in a completely overestimated itinerary travel time. An interesting method for solving these complexities is presented in Chap. 5.

In contrast, open toll highways do not need to reidentify the vehicle to charge the toll. Every vehicle is charged the same toll, resulting from an average trip in the highway. In this case, toll plazas are strategically located in the main highway trunk every now and then. The entrance and exit time complications do not appear in this case, but the exhaustive travel time measurement is lost. Note that in order to reidentify vehicles in this open configuration it is needed that the vehicles travel across, at least, two toll plazas, and leave a trace at each payment. This is achieved in case of electronic payment (e.g. credit card or electronic toll collection tag— ETC) and long trips. Short trips or cash users will travel unidentified. This will not be a major drawback, because for example approximately 80 % of toll highway users in European turnpikes are actually using electronic payment methods, which should be enough to obtain a representative average of the travel time measurement. In addition, in order to obtain the measurements in real time, the communications requirements are more challenging, as there is no baton travelling with the vehicle, and cross analysis of data of different toll stations is needed. Finally note that main toll plazas are usually located several kilometers apart. Probably, a finer discretization of travel time measurements will be desired. This can be achieved by installing ad hoc overhead gantries capable of reading the electronic toll collection tags "on route", see Fig. 1.4. This technology has been proved capable of identifying all ETC equipped vehicles even if they are travelling simultaneously in different lanes at speeds as high as 180 km/h. The main issue here is the penetration rate of these electronic tags, which nowadays is around 20 %, as a result of marketing policies (e.g. free dissemination of tags for frequent users) and priority benefits at toll plazas, where usually, equipped vehicles cross undisturbed (Table 1.1).

1.2.1.2 Vehicle Tracking

The second group of techniques for the direct travel time measurement is related to the vehicle tracking concept. In this case vehicles act as probes and record their position every defined time interval. There are not control points or infrastructure related monitoring equipment. The vehicles become active sensors, instead of being passive as in the previous case, and compute travel times by continuously tracking its trajectory.

Historically the vehicles used as probes have been dedicated cars. These probe cars traveled with the only purpose of gathering travel time data. This is the case of

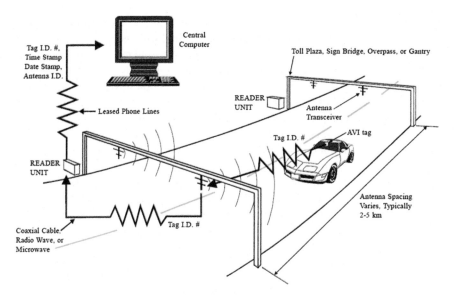

Fig. 1.4 Travel time estimation from ETC reidentification. *Source* Turner et al. (1998)

Table 1.1 Direct travel time measurement in toll highways: open versus closed toll configurations

Closed toll configuration	Open toll configuration
Vehicle reidentification by means of Toll Ticket (real or virtual)	Vehicle reidentification by means of credit card number or ETC identification
Exhaustive sample	Sample is made up only by electronic payment users who travel across two or more toll plazas
Toll ticket travels with the vehicle	Communication system between toll plazas needed for pairing
All on/off ramps are control points	Main trunk toll plazas are the control points. Additional ad hoc identification gantries may be needed
Travel times affected by entrance/exit times	Main trunk travel times are measured

traditional probe car data. In order to obtain a continuous flow of travel time measurements to be used as a real time information system in a highway corridor, the amount of ad hoc probe cars would be huge (e.g. probe cars at 3 min headways), and not sustainable in the long term. This traditional method is restricted to case specific studies.

The development of ITS (Intelligent Transportation Systems) and the popularization of GPS (Global Positioning System) technologies has favored that each vehicle which travels in a particular road could be a potential probe vehicle. These GPS equipped vehicles are nowadays regular transportation fleets (like buses,

1.2 Travel Time Measurement

parcel companies vans, roadside assistance vehicles, patrol service vehicles, taxi cabs, ...) which travel regularly over a selected route and who have at their disposal an active management center to elaborate the necessary data treatment. Take into account that the specificities of these fleets (e.g. heavy vehicles) may bias the sample. Currently, the weak point of the system is the data location transmission from the vehicle to the control center, usually using radio channels (e.g. GPRS system). These schemes have been expanded to every particular car who volunteers (this will solve the privacy issues). This extensive and automatic version of traditional probe cars needs a high penetration of on board GPS devices plus the collaboration of the driver in order to transmit the data. This is possible with the popularization of GPS-enabled smartphones (Herrera et al. 2010) (Table 1.2).

Table 1.2 Direct travel time measurement methods: benefits and drawbacks

		Benefits	Drawbacks
AVI methods	License plate, toll ticket, digital imaging, electromagnetic signature, Bluetooth devices	• Potential exhaustive sample (license plates, toll tickets) or high penetration rates (ETC devices) • Continuous flow of travel time measurements • Promising reidentification methods based on loop detector data	• Site specific and costly. The highway must be equipped with technologically advanced detectors • Clock synchronization between control points • Partial travel measurements in between control points are not available • Privacy issues • Difficulties in outlier detection in case of few measurements
Vehicle tracking	Traditional probe cars, GPS tracking, cell phones geo-location	• Not infrastructure related • Not spatial captive. The travel time measurement can be obtained between any two desired points • GPS equipped smart phones imply new opportunities for the method	• Traditional probe cars only useful for specific studies, but not for a systematic implementation • When based on specific fleets, may imply small and biased samples • Needs high penetration of GPS equipped vehicles • Large amount of data must be transmitted from vehicles to a data management center

1.2.2 Indirect Travel Time Estimation

Indirect travel time estimation is based in the measurement of fundamental traffic flow variables (flow, speed and density) in a particular spot of a highway and the extrapolation to a whole section, using some type of algorithm. Loop detectors are, by far, the most widely spread technology to collect flow, speed and occupancy (the proxy for the traffic density) of a traffic stream. Take into account that single loop detectors only collect flow, and occupancy, while speed must be approximated by usually assuming an average constant vehicle length. Besides, dual loop detectors are capable of measuring all traffic variables (i.e. flow, speed and occupancy).

Using loop detector data always imply the problem of data quality. Flow, speed and occupancy measured on a highway spot over short aggregation periods (of the order of some few minutes) suffer from important fluctuations, particularly when measuring instable stop and go traffic. Huge variations are possible in short time periods, and still the measurements are correct. This makes extremely difficult to detect erroneous loop detector data on real time, unless very abnormal data is measured. In contrast, smoothing or aggregating data over longer time periods (e.g. one day) makes the detector malfunctioning a lot more evident and easily detectable (Chen et al. 2003). Fortunately, because loop errors do not arise randomly in between correct measurements, this is enough in most cases (although several hours of undetectable malfunctioning may be unavoidable). Usually, detector failures respond to the breaking down of some part of the detector, and not to a circumstantial malfunctioning. This means that erroneous data usually respond to "stuck" measurements during long periods of time (days, weeks, months or even years) until the detector is repaired.

Loop detectors are an old technology, which require intensive and costly maintenance as they are exposed to severe conditions (i.e. traffic, extreme hot, extreme cold, water, road works ...). If this maintenance work is overlooked, frequent breakdowns occur. This implies great holes in the database, with the consequent implications in the algorithms relying on these data. These algorithms should be prepared to deal with this frequent missing data. Travel time algorithms rely on sectional measures. This means that, in general, no lane specific data is used, but data aggregated over all lanes. Detectors in different lanes can be considered as redundant (this is a simplistic approach, because traffic each lane has its own characteristics, and some analysis or applications need to measure this specific lane behavior). This implies that, in case some detector in the section is not functioning, it is easier to reconstruct the data of the whole section by using neighboring detectors, if they are properly working. Therefore, a complete failure only happens when all the detectors in the measurement section fail all together. This is not as rare as it may seem, because it is only necessary the failure of the roadside unit which steers the measurements of all the detectors in the section.

From the previous paragraphs must be concluded that data quality assessment and data reconstruction processes are the necessary first step in the utilization of loop detector data for whichever desired objective.

For the particular objective of indirect travel time estimation from standard loop detector data, two basic methodologies can be distinguished: the estimation from point speed measurements and the estimation from cumulative count curves. Both are analyzed in the following sections.

1.2.2.1 Indirect Travel Time Estimation from Point Speed Measurement

The first and most widely used approach for estimating travel times from loop detector data is the spot speed algorithm. This method is based in the extrapolation of the point speed measurement at the loop location to a complete freeway section The hypothesis considered in the application of this algorithm are that punctually measured traffic stream characteristics are representative of the whole assigned section.

First of all, this method relies on a speed measurement, which therefore needs to be accurate. Single loop detectors' approximations to speed are not enough, as the assumption of constant average vehicle length does not provide the necessary accuracy. In order to solve this problem, traffic agencies have tended to install detectors in a double loop configuration (i.e. speed traps) to accurately measure the vehicles' speed. In addition, in order to estimate travel times from averaged speeds, resulting from measured individual speeds over a time period in a particular spot of the highway, this averaging must report the space—mean speeds. The problem here is that the standard loop detector data treatment reports the time—mean speeds, and the raw data useful to compute space—mean speeds is eliminated. In this situation the obtained travel times with the proposed method would be generally underestimated.

In relation to the spatial representativeness of punctual measurements, different agencies use different speed interpolation methods between detectors (e.g. constant, linear, quadratic ...) trying to better approximate the traffic conditions in the stretch, but without taking into account traffic dynamics. As it is proved in Chap. 3, all of them are simplistic and inaccurate in congested conditions. In view of these limitations, and in order to obtain meaningful travel time estimation using these methods, detector density must be extremely high. This has forced a process of continuous increase in the loop detector density. While one single detector in between junctions was enough to measure average daily traffic (ADT) volumes, at least one double detector every 500 m is necessary to compute accurate travel times using these methods (see Fig. 1.5). Obviously this is not economically feasible for the whole highway network, and can only be achieved in some privileged stretches of metropolitan freeways. In conclusion, there are lots of kilometers of interurban highways with low surveillance density (e.g. typically one detector per section between junctions to fulfill the ADT requirements) where the systematic travel time measurement using the existing equipment is devoid of an adequate method.

Finally, it is worth mentioning that itinerary travel times are frequently obtained by the addition of several section travel times (where each of these sections is

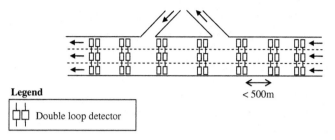

Fig. 1.5 Spot speed algorithm required surveillance configuration

defined by two loop detectors). Each section travel time is obtained from the average speed measured over the last few minutes (as a frequent update is desired). This means that the obtained itinerary travel time is not trajectory related. It is a like a picture of the actual travel times on the stretch. It is possible that none vehicle trajectory responds to this travel time. This temporal alignment concept of travel time is frequently known as ITT (Instantaneous Travel Time) and it may be considered to be the best approximation to the desired real time "future" information, without falling under the uncertainties of forecasting. More details regarding travel time definitions are presented in the next chapter.

1.2.2.2 Indirect Travel Time Estimation from Cumulative Count Curves

The alternative to avoid the required high surveillance density and the lack of accuracy of the spot speed algorithm in congested situations relies on a cumulative count balance algorithm, which estimates travel times directly from loop detector count measurements. The algorithm uses the entrance and exit flows in the highway stretch to calculate the travel time using the conservation of vehicles' equation. In order to apply the proposed method, the monitoring of the section under analysis has to be "closed", in the sense that all the on/off ramps must be monitored, in addition to some main trunk loop detectors (e.g. typically one on every section between junctions) (see Fig. 1.6). Under these conditions the vehicle accumulation in the section can be computed.

Despite the apparent potential and simplicity of the method, it is hardly used in practice. This may result from the oversight of the researchers' community to the practical problems, which appear in the implementation of the method. For instance, from the author knowledge, contributions are not found in the literature analyzing in detail the problematic detector drift phenomenon, which accumulate in the input/output curves until they become meaningless. The effects of inner section input/output flows at junctions are not treated either. These issues remain for further research.

1.2 Travel Time Measurement

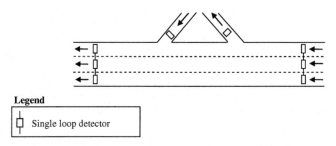

Fig. 1.6 Cumulative flow balance algorithm required surveillance configuration

1.2.2.3 Loop Detectors and Travel Times: Summary

The main conclusion of this section is that there is only one benefit of using loop detector data to compute travel times. This is, that loop detectors are already installed out there, and the marginal cost of this new application is small. And of course this is a major benefit. The rest are problems, which different methods suffer more or less. Table 1.3 summarizes the different problems that affect the different methods. This means that loop detectors usage as travel time estimation equipment is result of reminiscence of the past and given the inertia of the already installed equipment. Therefore, in new specific projects for systematic travel time estimation, where prevailing equipment is null, traditional loop detectors should not be considered as the best option.

Surely, in the curse of time, loop detector technology will be modernized. It is a perception of the author that given the structure of traffic monitoring systems and the interests of worldwide traffic agencies, the new detectors which finally will beat traditional loops must be capable of doing (at least) exactly the same functions which old ones do (count, occupancy and speed for all vehicles). They will be installed at exactly the same spots. So that from the traffic management center point of view in the daily life nothing will be changed. But of course the new detectors will be cheap, will benefit from a long lifespan, will be less intrusive into the pavement, will require low maintenance, will communicate wireless, will not need a power supply, and will be less prone to malfunctioning. The MeMS detectors developed by UC Berkeley engineers are a good example (Hill et al. 2000). This will be the time to incorporate to these detectors more advanced features, like reidentification capabilities.

In addition to Table 1.3 contents, there is another problem which affects all methods. This is the frequent malfunctioning of loop detectors, which is translated into empty holes in the database.

Being fair, there is also one benefit of loop detector data, and this is the instantaneity. This property, which can also be obtained with the direct tracking of vehicles, but not in the AVI methods, refers to the ability of the measurement equipment of providing an instant picture of what is happening, avoiding information delays.

Table 1.3 Methods of travel time estimation from loop detectors: benefits and drawbacks

	Benefits	Drawbacks
Reidentification of vehicles (electromagnetic signature, characteristic length or platoon structure)	• Spatial measurement (direct travel time measurement) • Active research with promising results	• Needs detector retrofitting (new hardware) • In case of in between junctions, in order to avoid false reidentifications, only rare vehicles are identified. Biased measurement in free flow, but not a serious flaw in congested conditions • Platoon structure is lost in congestion • Clock sincronization between detectors
Spot speed methods	• Instant travel time measurement • Intuitive algorithm • Directly applicable with current detectors in almost any configuration • Intelligent smoothing helps in improving the estimations	• For accurate measurement needs intensive (and costly) monitoring (i.e. one detector every 500 m) • For accurate measurement needs the computation of space mean speeds • Even fulfilling the previous conditions, travel time estimations will be flawed in congestion conditions
Cumulative count curve method	• Spatial measurement (it is like a direct measurement under FIFO assumption) • Instant and trajectory based travel time measurements available • Directly applicable with current detector technology	• Needs closed configuration of the monitoring (all input/outputs being monitored) • Detector drift is a major problem • Methods need to account for inner section input/output flows • Care must be taken with the necessary frequent reset

1.3 Data Fusion and Travel Time Forecasting

The word "forecasting" has connotations of uncertainty, chance, fate, or even other metaphysic implications which are far beyond human knowledge. These connotations have been transferred to traffic information (have you ever wondered why traffic information bulletins are usually grouped with weather forecasts? Like if both were uncontrollable natural forces).

It is obvious that previously to face the challenging problem of forecasting, the ability of measuring the objective variable must be mastered. It is reflected in the

1.3 Data Fusion and Travel Time Forecasting

present monograph that this is not the case if we consider highway travel times and the currently installed surveillance equipment. In general, all forecasting methods need to base their predictions on measurements. Therefore all scientific approaches to travel time measurement are necessary, even for forecasting, and play an important role in research evolution. This does not mean that research devoted to travel time forecasting is in the wrong track. It only means that in order to obtain the maximum benefits of this research and become fully applicable in most of the highway network, research on travel time measurement is equally necessary.

This book is devoted to travel time measurement and the usefulness of this information for real time traffic information systems. However, one must realize that a real time travel time information system also needs to predict over the very short term. This fact justifies the inclusion of the present section of the book.

When a driver enters a highway he would desire being told how long it will take his trip. This could be materialized in a futuristic example, as an individualized message from the radio in the car saying *"The travel time to your destination will be ... minutes"*. In reality this is a short term forecast. As it is a forecast, and future is always uncertain, a modification of the previous message to account for this fact could be *"The expected travel time to your destination is ... minutes"*. The word "expected" adds the idea of uncertainty in the information, and it means that this is the better information it can be provided, taking into account the actual traffic conditions and the typical (or recurrent) evolution in similar situations. In fact, the previous message could be further modified to quantify this uncertainty. An option could be *"There is a 90 % probability for the travel time to your destination being between ... and ... minutes"*. The confidence level could be avoided for marketing options, but the confidence interval could be kept as important information. One hundred percent confidence in the information cannot be guaranteed and still provide informative confidence intervals. This is due to the fact that non typical behavior may arise or non-recurrent events can happen (e.g. vehicle breakdowns, accidents...). By definition, non-recurrent events are those that cannot be anticipated, and therefore can only be taken into account once they have happened and become actual information. Despite all these uncertainties, what actually could and should (at least) be told to the driver is *"The current travel time to your destination is ... minutes"*. This is not his desired information, but at least it is certain (it corresponds to the last travel time measurement), and by sure it is better than unreliable predictions.

What can be concluded from the previous discussion is:

- Last updated travel time measurement is the best information that can disseminated on real time without falling in the uncertainties of forecasting. For long trips this can be significantly different than the travel time which finally experiments the driver. Frequent update of the measurements helps in informing on the evolution.
- Forecasting is uncertain. Quantification of the uncertainties should be provided.
- Forecasting is based on typical recurrent conditions. Non recurrent events will always remain as unpredictable, until they happen and can be measured.

Therefore a forecasting method must also consider actual traffic conditions (measurement) to account for non-recurrences which have already happened.

In view of these comments it seems clear that the key element to be considered in travel time forecasting is the horizon of the forecast. Two temporal horizons can clearly be distinguished: long term and short term forecasting. Long term forecasting refers to the fact that the instant of interest (horizon of the forecast) is far enough in time so that the current traffic conditions do not affect the forecasted travel time. The effects of all the non recurrent events which are currently happening will be vanished by the prediction time. In this long term forecasting, historical information of similar time periods is the basis for an accurate forecast. As an order of magnitude, it could be considered that long term horizons start from the next day, from the instant of making the prevision. Besides, short term forecasting refers to a forecasting horizon where current traffic conditions prevail. The current measured travel times, already affected by the present non-recurrences are by far more informative than historical information. This horizon has a magnitude of the next hour. Obviously this classification is neither strict nor discrete, and in between a continuous gradation of medium term forecasts, where both, current and historical information matter, can be found.

Different forecasting horizons respond to different objectives. For instance long term forecasting provide a useful trip planning tool for users and may help traffic agencies in setting in advance some operational schemes. Short term forecasting is, as it has been stated, useful for real time travel time information, where the horizon of the prevision is precisely this future travel time estimation. Wide range of medium term horizons may be useful for traffic agencies to assess the real time decision making process in setting different operational schemes on the highway.

Depending on the forecast horizon different methodologies could be applied. Each technique is suitable for a particular horizon and there is no methodology to foresee the traffic conditions in all the horizons. For instance, long term prediction is usually based on statistical methods where the main problem to solve lies in determining which similar episodes should be grouped (Chrobok et al. 2004; Danech-Pajouh 2003; Van Iseghem 1999) as a unique behaviors. The cluster analysis method represents a useful statistical tool to accomplish this objective as presented in Soriguera et al. (2008) and Rosas et al. (2008). The results of this analysis consist in a year calendar representing the grouping of different types of days. Note that this statistical analysis can be directly applied to a travel time database (provided it exists from a systematic measurement over some years, which is still very rare), in which case, travel time patterns (with its mean and percentiles) would be directly obtained. The alternative could be to apply this pattern characterization to the traffic demand (origin—destination matrixes) if this is the more common available information. Then, a traffic model would be needed to translate these O-D patterns into travel time patterns.

Several benefits appear if using this modeling approach. First of all, the demand is related to the travel behavior of people. Although the behavior of individuals may be irrational, on an aggregated scale the behavior of the demand is rational and with a smooth evolution. Therefore it should be easier to predict than the volatile travel times, which are a derivative variable from the traffic demands. Travel times are characterized by a constant free flow travel time for a wide range of traffic conditions with extremely sensible increases when the demand reaches the congestion threshold. In those situations little demand variations may imply significant travel time changes. Therefore, travel times must be more difficult to predict directly.

An additional benefit appear in the medium term forecast, where in addition to the historical information, non-recurrent or rare events which are actually taking place must be considered (e.g. road works, lane closures, accidents, bad weather...). Surely travel times in these situations could also be found in the database, but their statistical significance would be very low and the number of possible combinations may result prohibitive. As the demand would remain unaffected (or at least less affected), these non-recurrences are more easily taken into by modeling. A similar situation appears in case of a modification of the infrastructure (e.g. the elimination of a severe bottleneck). In this case the historical series of travel times will be lost. However the modifications in the modeling approach would be slight. The difficulties, like in all the modeling approaches, may appear by the excess of calibration parameters, in addition to the associated modeling errors.

In relation to the short term forecasting, several methods are under research. Some of them are based in spectral analysis, autoregressive time series analysis or in Kalman filtering (Clark et al. 1993). None of them seems to provide enough accuracy, and errors around 25 % are reported in Blue et al. (1994). More recently the research interest is focused on neural network models and artificial intelligence (Dia 2001; Dougherty and Cobbett 1997). Between them, data fusion schemes also appear as an alternative. This approach is treated in more detail in the next section, as it is the approach used in Chap. 6 contribution.

1.3.1 Data Fusion Schemes

Multiple source Data Fusion (DF) consists in the combined use of multidisciplinary techniques, analogous to the cognitive process in humans, with the objective of reaching a conclusion in relation to some aspect of reality which allows taking a decision. In terms of the accuracy of estimation, the combined use of data from multiple sources makes possible to achieve inferences, which will be more efficient and potentially more accurate than if they were achieved by means of a single source. Data fusion schemes are widely used, mainly in digital image recognition or in the medical diagnosis. During the last decade it has also been applied with data related to the transportation field (Hall and McMullen 2004).

In the next section, a conceptual classification of the different data fusion techniques found in the literature is presented.

1.3.1.1 Behavioral Classification of Data Fusion Operators

This section aims to describe the different behaviors that a fusion operator may have. First of all, it is necessary to define the following notation:

- "x_i" is a variable which represents the credibility associated to a particular data source "i". "x_i" takes values between 0 and 1.
- "$F(x_1, \ldots, x_n)$" is the credibility resulting from the data fusion operator. It also takes values between 0 and 1.

In order to simplify the concepts, assume that two types of information are going to be fused. In this case "$x_i = (x, y)$". In this situation and according to Bloch (1996), data fusion techniques can be classified, in relation to their behavior as severe, cautious and indulgent:

1. *Severe*: a fusion operator is considered severe if performs with a conjunctive behavior. This is:

$$F(x,y) \leq \min(x,y) \qquad (1.1)$$

2. *Cautious*: a fusion operator is considered cautious if it behaves like a compromise. This is:

$$\min(x,y) \leq F(x,y) \leq \max(x,y) \qquad (1.2)$$

3. *Indulgent*: a fusion operator is considered indulgent if performs with a disjunctive behavior. This is:

$$F(x,y) \geq \max(x,y) \qquad (1.3)$$

1.3.1.2 Contextual Classification of Data Fusion Operations

This section classifies the fusion operators in terms of their behavior with respect to the particular values of the information to be combined, and to the use of other external information available: the context.

1. *Context Independent Constant Behavior Operators (CICB)*: This class of operators it is constituted by those operators with the same behavior independently of the particular values of the information to fuse. In addition, the data fusion does not take into account any other external information (apart from the values to fuse) regarding the context of the fusion. The operator will have always with the same behavior: severe, cautious or indulgent, whichever, but always the same. The most famous techniques in this class are the Bayesian fusion or the Dempster-Shafer technique.
2. *Context Independent Variable Behavior Operators (CIVB)*: This second class of operators groups those operators which, being independent from the context of

the fusion, they depend on the values of "*x*" and "*y*". Therefore, its behavior may change depending on the values of the variables to fuse. Examples of this class are the artificial intelligence or the expert systems.
3. *Context Dependent Operators* (*CD*): the behavior of this class of operators depends not only on the value of the variables to fuse, but also on the global knowledge of the context of the fusion (e.g. knowledge of the credibility of different sources in different situations). Some applications of the fuzzy sets technique are CD operators.

1.3.1.3 Mathematical Classification of Data Fusion Operations

In accordance to Hall and McMullen (2004), data fusion techniques can also be grouped considering the mathematical logic used to take into account the lack of credibility of data. For instance this is the main difference between Bayesian, Dempster—Shafer and fuzzy sets techniques, while artificial intelligence methods are differentiated by their learning process.

1. *Probabilistic logic*: It is the most widespread mathematical logic, with robust solid mathematical foundations given by the classic probability theory. They need an empirical construction of the probability density functions and of the conditional probabilities. These may impose severe restrictions to the method in complex problems.
2. *Evidential logic*: Mainly represented by the evidential theory of Dempster—Shafer, which allows to include the confidence given to the probability of a particular event. It is useful in those situations where the probability density functions cannot be considered as correctly measured, but only approximated. A level of credibility is given to these functions and the ignorance can be explicitly considered.
3. *Fuzzy logic*: Fuzzy logic, derived from the fuzzy sets theory firstly developed in 1965 by Lotfi Zadeh at UC Berkeley, still today is a highly controversial theory. In fuzzy sets theory, the belonging to a set property is represented by a value between 0 and 1. Equivalently, in fuzzy logic the veracity of an assertion can also vary between 0 and 1, and it is not limited to true or false as in the bivalent logics. In this sense, fuzzy logic is multivalent.

1.3.2 Bayesian Data Fusion

A little more attention is given to Bayesian data fusion schemes as it is the one used in Chap. 6. This fusion technique, based on the Bayes' Theorem of classical probability theory, belongs to the class of algorithms which use a priori knowledge of the observed variables in order to infer decisions on the identity of the objects being analyzed. The Bayesian method provides a model to compute the posteriori probability of a given context.

Analytically, the Bayesian data fusion method can be formulated as follows. If "E" is the object to evaluate and "x_1", "x_2" the information elements obtained from two sensors, from the Bayes theorem it can be stated:

$$p(E|x_1,x_2) = \frac{p(E,x_1,x_2)}{p(x_1,x_2)} = \frac{p(x_2|E,x_1) \cdot p(E,x_1)}{p(x_1,x_2)}$$
$$= \frac{p(x_2|E,x_1) \cdot p(x_1|E) \cdot p(E)}{p(x_1,x_2)} \quad (1.4)$$

Assuming independence between the measurements of different data sources, Eq. 1.4 can be simplified to:

$$p(E|x_1,x_2) = \frac{p(x_2|E) \cdot p(x_1|E) \cdot p(E)}{p(x_1) \cdot p(x_2)} \quad (1.5)$$

Generalizing the method for "n" sources of information, we obtain:

$$p(E|x_1,\ldots,x_n) = \frac{p(x_n|E,x_1,\ldots,x_{n-1})\cdots p(x_1|E) \cdot p(E)}{p(x_1,\ldots,x_n)} \quad (1.6)$$

Again, assuming independence between data sources:

$$p(E|x_1,\ldots,x_n) = p(E)\frac{\prod_{i=1}^{n} p(x_i|E)}{\prod_{i=1}^{n} p(x_i)} \quad (1.7)$$

Finally, if "$\Omega = \{E_1,\ldots,E_r\}$" is defined as the set of "r" possible states, the final decision may be reached according to the following criteria:

Maximum a posteriori probability rule: The most probable state is the one with higher a posteriori probability.

$$E_k = \arg\max_{1 \le i \le r} \{p(E_i|x_1,\ldots,x_n)\} \quad (1.8)$$

Maximum likelihood rule: The most probable state is the one with a higher value in the likelihood function.

$$E_k = \arg\max_{1 \le i \le r} \left\{\prod_{j=1}^{n} p(x_j|E_i)\right\} \quad (1.9)$$

1.3 Data Fusion and Travel Time Forecasting

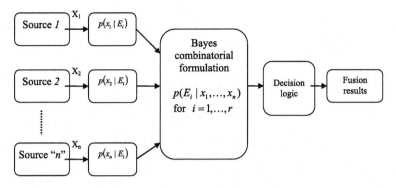

Fig. 1.7 Bayesian data fusion method

Both decision rules converge to the same decision when the a priori probabilities are uniform, "$p(E_i) = 1/r$".

Therefore, the Bayesian data fusion method is a severe technique where the fusion operator is independent on the context and on the value of the variables to fuse. The main advantage of the method is a solid mathematical background, where credibility is formulated as a probability function. Using this technique, the a priori knowledge can be expressed in stochastic terms in order to obtain the most probable state of the system.

In practice, it is needed to obtain the conditional probability functions, "$p(x_j|E_i)$" and the a priori probability functions, "$p(E_i)$" which model the contribution to the final knowledge of each source. It is a common assumption when there is a completely ignorance to consider uniform a priori probabilities, "$p(E_i) = 1/r$", then the a priori knowledge does not contribute to the final decision, and estimate "$p(x_j|E_i)$" from a statistical learning method (Fig. 1.7).

1.3.3 Main Benefits and Drawbacks of Data Fusion Schemes

Before deciding on the application of data fusion techniques it is interesting to know the overall expected benefits and the potential drawbacks in the application. Results obtained in Nahum and Pokoski (1980) suggest (see Fig. 1.8):

- Combining low credibility sources of information (low probability of accurate measurement of single data sources "$P_N < 0.5$") does not contribute in a better final estimation. Therefore an initial requirement to obtain some benefit from data fusion is a minimum precision of data sources.
- Combining high precision data sources ("$P_N > 0.95$") do not contribute in significant benefits. If sensors are precise, there is no point in working for further accuracy.

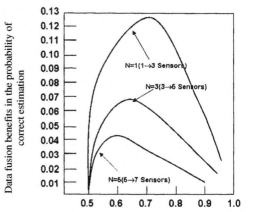

Fig. 1.8 Expected benefits of data fusion in relation to the number and accuracy of sensors. *Source* Nahum and Pokoski (1980)

- Increasing the number of sensors, increases the benefits of the fusion. However the marginal benefits decrease with the increase in the number of sensors.

In this context, the main advantages of the data fusion schemes are:

- Operational robustness, as one sensor may contribute in the final estimation while the others are inoperative or only partially operative. This contributes in a better temporal coverage of the estimations.
- Increase of the spatial coverage of the measurements, as different sensors can be installed in different spots with different spatial coverage.
- Reliability increase, as the veracity of the information is contrasted by redundant data sources.

The main drawbacks of data fusion may be summarized as:

- Need for a minimum number and a minimum quality of the contributing data sources. This aspect steers the results of the fusion.
- Previous knowledge of the quality of the data provided by each type of sensor. This, although not being restrictive, it may improve significantly the quality of the results.
- There is not a "perfect" data fusion operator. Each algorism has its own strengths and weaknesses.
- Common lack of training data, necessary for the statistical learning of the algorithms.
- Dynamic process. It is difficult to evaluate the results, as the efficiency of the method will improve gradually with the learning of the method.

Finally, it is interesting to stress the computational effort implied by the data fusion scheme in relation to the whole computational effort required by a data information system. According to Hall and McMullen (2004) it can be estimated in an 11 % (Fig. 1.9).

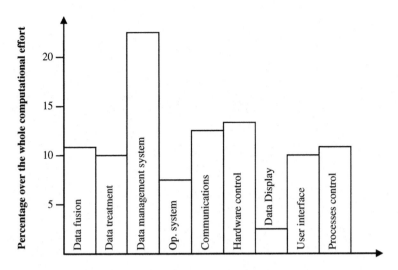

Fig. 1.9 Data fusion computational effort. *Source* Hall and McMullen (2004)

1.4 Travel Time Information Dissemination

A fundamental component of a travel time information system is the dissemination of the information. Dissemination techniques can be clearly divided into two main types: pre-trip information and on-trip information. Pre-trip information allows trip planning while on-trip information enables network users to (possibly) modify the initial planning according to current traffic conditions.

To be effective, traffic information must be short, concise, quantified and specifically addressed to the receptor. Travel time information itself fulfills the three first conditions, and the dissemination technology must fulfill the last one. That is, the travel time information to be conveyed must be that of specific interest to the driver.

There are several techniques for travel time dissemination, each one related to a particular technology. Their detailed characteristics are shown in Table 1.4.

1.4.1 Information Before Departure

Pre-trip information allows the user to decide the time of travel, the mode of travel, or even cancel the trip all together. Pre-trip information reduces the risk of delivering goods late or arriving late at the destination in general.

There are many options available to disseminate pre-trip information. Traditionally, newspaper or radio has been a source of traffic information, especially in case of a special event such as a sporting event or a festival. These types of

Table 1.4 Travel time dissemination techniques

Techniques	Characteristics
Radio broadcasts	• Traffic information bulletins • Capable of disseminating pre-trip and on-trip information • No user discrimination. Each driver must carefully listen to the whole bulletin and select his own information of interest • Discrete information times, subject to the scheduled bulletins • In case of short range dedicated radio signal, this last two limitations can be overcome
TV Broadcasts	• Only pre-trip information • No user discrimination • Discrete information times
Press	• Only pre-trip information • No user discrimination • Discrete information times
Traffic Call Center	• Capable of disseminating pre-trip and on-trip information, with the limitation of on-trip telephone calls • It is a service on demand. Driver must ask for it and usually pay a price for it. This implies a limitation of access to the information
Variable message signs	• Capable of disseminating pre-trip and on-trip information • Specifically addressed to the driver, as only inform the drivers who travel below them • Continuous and very accessible information
Internet	• Only pre-trip information (on-trip using smart mobile phones) • It is a service on demand. User must log-in and ask for a specific itinerary
Car navigator (RDS-TMC radio signal)	• Capable of disseminating pre-trip and on-trip information • Specifically addressed to the driver (GPS/GSM/UMTS) • Continuous and very accessible information
Cellular phone (text service)	• Equivalent to a call center with the improvement that you can subscribe to a particular corridor and receive information without asking every time for it
Information points	• Capable of disseminating all types of information and discriminate between users. However its accessibility is very low because the driver must stop at the service area to obtain the information

information basically warn network users of the possible delays and provide information on the extent of the disturbance on the network. Even though these types of traditional measures might be considered as too static, they may provide valuable information to the users of the network and mitigate possible unreliability impacts, if correctly targeted.

Nowadays, most of the dissemination techniques associated with pre-trip information are internet-based services and also provide up-to-date information into mobile phones. A number of service companies offer calculation of journey times or travel information with added value. The first websites grew up in the mid 1990s. Most of them originally provided traffic information for a certain region but recently they have been extended to cover whole network. These websites initially targeted the general public but then began to offer professional solutions such as geographical location of clients.

1.4.2 Information En Route

Using on-trip information may mitigate undesired impacts in a cost-effective way. Depending on the information, network users may decide to change their route, if alternative is available, still arriving on time at their destination. Users may also reduce the impact of arriving late by rescheduling their deliveries or planned activities and hence reduce the ripple or snowball effect of them being late. Even in the case there is no possibility to react, just the information of being late reduces the stress related to not knowing how long the possible delay may last.

Electronic information signs are now a familiar sight across the world on motorways and trunk road network. These signs are the main technology to provide on-trip information. The warn drivers of emergencies, incidents and road management. They are aimed at improving safety and minimizing the impact of congestion. Variable Message Signs (VMS) is a term often used to describe these signs. The main purpose of VMS is to communicate information and advice to drivers about emergencies, incidents and network management, aimed at improving safety and minimizing the impact of congestion. Messages displayed on VMS are often limited to those that help drivers complete their journey safely and efficiently. There are a number of types of VMS in use around the world and they provide the capability to display a wide range of warnings, messages and other traffic information.

The telephone is another way of transmitting on-trip travel time information to the driver. It should be one easy-to-remember number regardless of the traveler's location. The U.S. Department of Transportation (USDOT) petitioned the Federal Communications Commission to designate a nationwide three-digit telephone number for traveller information in 1999. This petition was formally supported by 17 State DOTs, 32 transit operators, and 23 Metropolitan Planning Organizations and local agencies. On July 21, 2000 the Federal Communications Commission designated "511" as the single traffic information telephone number to be made available to states and local jurisdictions across the country. An interesting point here is that the number is national but information is local.

Dynamic vehicle guidance and navigation services which include real time traffic data is at an early stage of development. However the integration of a status quo within the transport systems is essential to provide reliable travel time information. Further development is necessary in order to tap the full potential. This especially includes the cooperation of different players in the transport system, such as road administrations on various levels as well as public and private transport service providers. A major step would be the intermodal integration of guidance and navigation applications. Today capable techniques are mostly available; the cooperation between the different administrative levels needs to be improved. This would lead to integrated traffic information, which would enable also commuters—like combined traffic today—to choose the best available multimodal connection for the actual trip. Closely related car integrated systems for driver assistance will help to minimize the probability of incidents which influence reliability negatively.

Most of the applications currently are provided on commercial basis. Many navigator models, for example, already provide real-time information on incidents, weather, and traffic to mobile phones and car navigators. They calculate estimated travel times and take into account incidents in order to improve the estimated travel time. However, they are often limited by their capability to take into account changes in the traffic conditions due to new information in real-time. Most portable GPS devices today offer only a single route choice to the destination. This has one major drawback. In case of an incident, all drivers will follow the same advice given by the navigator. This will result congestion and delay on the new route. Some applications have recently emerged which provide alternative route options.

1.4.3 Comparison of Different Traffic Information Dissemination Technologies

Taking into account these factors, multicriteria analysis of different traffic information dissemination technologies has been performed (Table 1.5). It can be

Table 1.5 Multicriteria analysis of traffic information dissemination technologies

Mark	Technology	Description	Pre-Trip	On-Trip	Specifically addressed	On demand	User friendly
10	Car navigator	On vehicle device RDS-TMC	√	√	√	X	√
10	VMS	Variable message signs	√	√	√	X	√
8	Radio broadcasts	Radio bulletin	√	√	X	X	√
7.5	Cellular phone	Text services	√	√	√	√	=
7	Phone	Traffic call center	√	√	√	√	X
2.5	TV broadcasts	TV bulletin	√	X	X	X	√
2.5	Press	Conflictive days announcements	√	X	X	X	√
2.5	Internet	Online services	√	X	√	√	√
2	Information points	Service area information	√	√	√	√	XX

Note √ Good, = Medium, X Bad, XX Very bad

concluded that car navigators and VMS are the technologies with higher dissemination potentialities. These results are in accordance to the current practices. Moreover results also agree with user perceptions, as car navigation devices are currently bestselling car items.

1.5 Value of Travel Time Information

Funding information infrastructure is still a challenge. Traffic information is often organized by the state for national roads, by concessionaires on toll motorways, and departments and towns for their local road networks. Information equipment is generally financed by the various authorities responsible for their own networks.

Network users, for their part, have grown accustomed to considering that information should be provided free of charge. They see no reason to pay for access to information once they have paid for their car with all its accessories and paid their taxes, etc. In spite of this unwillingness to pay, there is abundant evidence that travelers place a high value on travel time information.

A study of motorists' preferences by Harder et al. (2005) found that travelers would be willing to pay up to $1.00 per trip for convenient and accurate travel-time predictions, such as when traffic is delayed which alternative routes would be faster. An interesting question that arises from this result is when travel time information has a great added value that will make drivers willing to pay for it?

The answer to this question should be when travel time information allows a benefit greater than their cost. A quantitative response to this question will not be provided here, and stands out as an interesting issue for further research. However it is evident that the value of travel time increases when the provided information is really informative.

This means that in case a freeway corridor is always free flowing, the free flow travel time information will be almost meaningless and without any value. Then, for travel time information being valuable it needs to vary (travel time variability)? Travel time will be informative when it varies, that is in congestion episodes? Not always. Note that if exactly the same congestion exists every day at exactly the same time, the travel time information will also tend to be non informative (at least for commuters). Note here that a first distinction appears: Travel time information may be valuable for a sporadic user, and meaningless for a commuter. It all depends on their baseline level of information, obtained from the experience.

In reality, what affects the value of travel time information is its ability to inform of unexpected travel times (larger, but also shorter). And here, the word unexpected matters. The value of travel time information is related to reducing the uncertainties. This concept is known as travel time reliability.

This unexpected behavior depends on two components: the baseline level of information of the user and the traffic characteristics of the highway. This means that in some infrastructures travel time information can be valuable while in others always meaningless for almost all drivers. It also means that the value of travel time information will not be the same for all drivers (also if the socioeconomic differences between drivers, which affect the willingness to pay, are not taken into account). These issues are analyzed in the present chapter.

The drivers' aversion to road travel time unreliability results from its high "costs". The costs of unreliability are due to two main reasons: arriving too late and arriving too early. Both situations imply an extension of the waiting time (on route or at destination), with the aggravating circumstance of loosing meetings, connections... the undesirable snowballing effect of arriving too late. To prevent the later, drivers allow extra time (buffer time) for the journey, increasing the probability of arriving too early. It should be clear that as travel time unreliability increase, so do the waiting times. Even on the unlikely situation of being on time, in unreliable road conditions, the anxiety and stress caused by the uncertainty in the decision-making about departure time and route choice imply an additional "cost" for the driver. The costs of unreliability seem to be clear.

The value of travel time information is directly related to its ability of reducing unreliability and its associated costs. As it will be explained next, travel time information by itself can mitigate the unreliability and its consequences. The information does not stop an incident happening but rather reduces the costs that arise from the incident.

1.5.1 Travel Time: Variability, Reliability and Value of Information

On one hand, travel time reliability can be defined as the lack of unexpected delays in a road stretch. Then an itinerary could be considered as reliable, in terms of travel time, when the actual travel time of a particular vehicle is close to its expected travel time. Take into account that the expected travel time may include the expected recurrent delays. On the other hand, travel time variability can be defined as the variation in travel time on the same trip traveled in different times of the day or in a different day in the week.

Note that, with these definitions, travel time reliability depends on the driver's expected travel time. This expectation varies with the driver's information, which could be result of experience gained from past trips or directly provided by the road operator.

In the case of a sporadic driver with no travel time information from the operator, it is probable that his knowledge is limited to the free flow travel time (from the expected average speed for the type of road and the distance). Note that in this situation (see Fig. 1.10a) the travel time unreliability is very high if traveling in a heavy traffic freeway stretch with its associated travel time variability. This results from a very wide travel time frequency distribution with a high range of possible travel times.

1.5 Value of Travel Time Information

If the knowledge of the driver improves, due to his acquired experience on the corridor or due to travel time information provided by the operator, and he knows if traveling in peak or non-peak hour and the associated expectations for the travel time (e.g. daily commuter with knowledge of recurrent delays—expected delays at same time of the day in some type of days), see Fig. 1.10b, the travel time unreliability is reduced due to the reduction of possible travel times in each traffic condition (congested or not). Finally, if the driver has a very good knowledge of the traffic conditions in his trip (very accurate travel time information on the freeway), different expectations of travel time within the congested period can also be

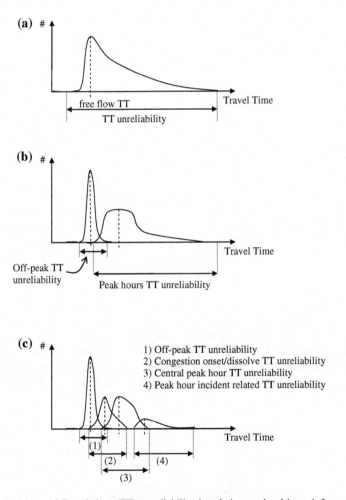

Fig. 1.10 Variation of Travel time (TT) unreliability in relation to the drivers information in a multilane freeway. **a** Information limited to free flow travel, **b** Different travel time expectations for peak, **c** Detailed travel time expectations for the congested situation

predicted (Fig. 1.10c). Then, travel time unreliability can be defined as the width of the frequency distributions of travel time around the driver expected average travel time and considering his a priori knowledge.

In this context the relationship between travel time variability and reliability is not direct, as a heavy peak hour freeway with high travel time variations within a day, could also be a very reliable freeway if accurate information is provided to the driver and an efficient incident management system is fully working in order to avoid non recurrent incidents. In contrast, a road stretch with less travel time variability could be very unreliable for the driver if no information is provided and frequent incidents imply serious unexpected delays.

The conclusion is clear, as higher is the information provided to the driver, closer is the expected travel time to the real travel time and so, higher is the reliability of the infrastructure. Therefore, information reduces the unreliability, but it will be when reduces unreliability costs that will become most valuable. As stated before, reducing the unreliability reduces stress, and this has a cost. Then, even not being able to modify any characteristic of the trip (i.e. the driver is trapped in the highway) if the information reduces unreliability it has a value: it reduces stress and makes rescheduling the events at destination possible, smoothing the ripple effect.

However, the value of information will be higher when making possible to reduce the other costs of unreliability (e.g. avoiding arriving late or early due to an excessive buffer time). To accomplish this objective and acquire value the travel time information has to reach the user enough in advance (pre-trip information) to be able to modify the instant of departure (to account for unexpected increases or decreases in travel times), or modify the mode or the route choice. Once on route, the instant of departure it is already decided. There only remains to act in the route or mode choice. Then, the location on the highway where on-trip travel time information is provided affects significantly its value. It must allow route choice (in advance of main junctions) or mode choice (before park and ride stations).

To finish this discussion, a note of the system wide effects of travel time information is pertinent. In a highway network where all drivers have perfect information, they will distribute over the different route options taking into account their own benefits. Considering the increase of the cost of traveling through a link with the increase of the demand (see Fig. 1.11), the system will reach a user equilibrium (recall the Wardrop (1952) principles on network equilibrium). This user equilibrium, which could be considered to be reached in a current day (with no

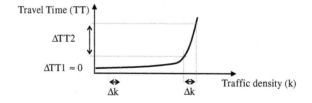

Fig. 1.11 Density versus travel time diagram

non-recurrent incidents) in a metropolitan highway network where most of the drivers are commuters with good knowledge of recurrent conditions, it does not have to correspond to the system optimum. With the universal dissemination of travel time information, the user equilibrium will also be reached in non-recurrent conditions. This means that if every driver is provided the same real time information of current traffic conditions, the system will evolve to user equilibrium where any driver by itself will not be able to improve his performance by switching routes. In this case the value of the information will be again very low. This means that the value of the travel time information diminishes with the number of drivers which have the same information. In general, the most valuable information is the one that few people know. In addition it is probable that the user equilibrium reached in case of universal information in non-recurrent conditions does not represent an optimum, or even the paradox case where the equilibrium reached is even worst, from a system wide point of view, than the original situation without information. The fact that information may not only re-route drivers but also divert them to other transportation networks (possibly less sensible to the demand) should be considered as a positive aspect of the information in this case.

This opens up a new research direction which should try to establish better strategies for the dissemination of the information with the objective to improve the system wide performance. It is probable that not all the drivers have to receive the same information, and here equity issues will appear.

1.5.2 Travel Time Unreliability Expected Behavior in Multilane Freeways

Figure 1.10 represents the approximate qualitative behavior of travel time frequency distributions for a heavy traffic multilane freeway. Some aspects that may be interesting for the reader are the following:

- Travel time distributions over long time periods (several hours) are highly skewed with a long right tail. This is due to the inclusion of different traffic states (Fig. 1.10a). As the time window is reduced, the travel time distributions tend towards a normal (Fig. 1.10b, c).
- Travel time unreliability in congested situations is significantly higher than in free flow situations (Fig. 1.10b, c). This results from the random behavior of traffic demand within a particular traffic flow pattern and from the great no linearity between traffic density and travel times (see Fig. 1.11). Note that in the congested situations, little variations in density produce high variations in travel time, while the same variation in density in the free flow zone, does not imply a meaningful change in travel times.
- For the same reason, catastrophic delays due to incidents in the freeway only occur in the heavy traffic periods, while very lower effects are expected in light traffic conditions.

1.5.3 Sources of Travel Time Unreliability: What Could Be Done?

As seen (Fig. 1.10c) even if the recurrent delays are known and provided to drivers by the highway operator, some unreliability remains. This could be considered as the "baseline" travel time unreliability.

Baseline unreliability, the source of value for travel time information, responds to two main phenomena: the random variation on transportation facilities demand and the probability of an incident in the freeway. It is shown in Fig. 1.11 that little variations on the demand for transportation between to similar days at the same time can produce considerable variations in travel time (in heavy traffic conditions). Additionally to this first source of unreliability it should be also taken into account the incident related unreliability.

For incident related unreliability it should be understood that "something" is happening in the freeway that implies additional delay beyond the usual random variations. These incidents can arise from two situations: an unusual increase in the demand (e.g. due to an especial event) or a reduction in the supply (i.e. capacity) of the freeway (e.g. vehicle breakdown, crash, road works, and bad weather). Note that bad weather sometimes can also imply an increase in the demand.

As stated previously, in general unreliability can be mitigated by information. In addition, each source of baseline unreliability must be faced in a different way. Table 1.6 present some possible strategic and operational measures to reduce the travel time unreliability in a freeway stretch. Note that each of the unreliability sources produces considerably more negative effects on travel times in heavy traffic conditions. If for some reasons (e.g. budget limitations) the mitigation measures can only be applied in reduced time windows, priority have to be given to congested periods with no doubt.

Table 1.6 Sources of travel time unreliability and possible mitigation measures

Source of TT unreliability	Possible mitigation measures
Recurrent demand variations	Pre-trip information in travel times
Random variation on transportation demand	Operational measures to lessen random variations (e.g. variable road pricing, ramp metering, dynamic flow control…)
Scheduled especial event	Information to road users in advance, provide information on alternative routes or alternative transportation modes, increase supply when possible by switching direction of some lanes
Vehicle crash or breakdown	Rapid response strategy. To be quick and effective in reaching the incident point and clearing the freeway
Road works	Scheduling strategy of the works taking into account traffic patterns and minimum capacity affection. (i.e. off-peak road works)
Bad weather	Difficult to mitigate this baseline unreliability. Increase of reliability can be achieved by means of information by obtaining specific patterns for bad weather conditions and accurate weather forecasts

1.5 Value of Travel Time Information

Baseline unreliability cannot be eliminated but should be limited. The question that remains for the policy makers is which should be the admissible unreliability threshold? And how bonus/malus economic incentives applied to operators could help to reach an admissible situation? But the answer is still far away. Note that a previous step is measuring unreliability, which means measure travel times (also a first mitigating measure), and as it has been stated in this monograph this is not a trivial task given the actual surveillance equipment.

1.5.4 Travel Time Information System: Levels of Application in a Road Network

The value of travel time information varies greatly with the variability and mainly unreliability of travel times on a particular stretch of road. The value that drivers give to travel time information is different if traveling in an unreliable stretch or in a very reliable one. In turn, this variability and reliability of travel times depends on the physical characteristics of the facility (number of lanes, slope, junctions ...) and on its relation to the demand for travelling through it.

Because always exist a budget limitation in the intensive monitoring of the network, and road networks are vast, priorities must be given to certain locations. The level of surveillance will directly affect the accuracy of the resulting travel time information. The basic criteria in the deployment of a travel time information system on a highway network should be the value of travel time information in each stretch and therefore the unreliability of travel times. Within the high travel time variability stretches (i.e. congested stretches) the value of the information provided by the road manager to the driver will be higher in those situations less predictable by driver experience in the corridor. This, points out the fact that for example in a freeway to reach the central business district of a big city, very congested every working day but in a similar magnitude, travel time information is less valuable due to the previous knowledge gained by commuters of the conditions in the freeway. In contrast, another facility also with heavy traffic near capacity, where the breakdown is less predictable and travel time for a particular departure time ranges from free flow travel time to severe and unexpected delays caused by frequent incidents, road works, bad weather or increased demand due to special events or seasonality of the facility, implies great value for travel time information provided by the operator. This means that the criteria for selecting priority corridors respect travel time information has to consider not only travel time variations in the whole day (level of congestion) but also the frequency of congestion (i.e. variation of travel times across days for a given departure time).

Finally, the last two indicators to take into account should be the demand of the corridor (hourly traffic volume in rush hours), and the monetary cost of implementing the surveillance equipment (dependent on the already existing equipment). From a cost-benefit analysis will result that priority must be given to those corridors where less money benefits more drivers (Table 1.7).

Table 1.7 Factors to take into account in a multicriteria analysis for selecting priority corridors to implement a travel time information system

Factor	Measures	Priority to
Congestion level for a particular type of day	• Travel Time Index: $TTI = \frac{Max\ TT\ rush}{Free\ Flow\ TT}$ • Peak Delay = Max TT rush − Free Flow TT	Higher TTI Greater delays
Frequency of congestion	• 90th percentile − median for the peak hour TT distribution • # days with congestion/total # of days within each group of types of day	Higher values of these unreliability measures
Average hourly traffic volume in the rush period	• Total volume of vehicles served in the rush period/duration of the rush	Higher volumes
Surveillance implementation cost	• Depends on the selected technology and the already existing equipment	Lower costs

Table 1.8 Priority corridors to implement a travel time information system in the Catalan network (northeast of Spain)

Priority	Corridors	Justification
1	Toll highways with high traffic volumes	• Severe congestion • Medium frequencies (high unreliability) • High traffic volumes • Existing tolling infrastructure
2	Freeways around main cities	• Severe daily congestion • High traffic volumes • Already existing intensive surveillance equipment
3	Seasonal corridors • Winter mountain corridors • Summer coastal corridors	• Severe sporadic congestion • low frequencies (high unreliability) • Low surveillance at the current time

These concepts were applied to the Catalan road network in Soriguera et al. (2006) taking into account the levels of existing surveillance, the frequency of congestion and the AADT (Annual Average Daily Traffic). Considering also the severity of this congestion, the rush hour durations, and the most suitable technology to measure travel time in each corridor, priorities are obtained, and can be seen in Table 1.8.

1.6 Conclusions and Further Research

It is possible to develop an accurate real-time travel time information system with the existing surveillance equipment. This sentence summarizes the research presented in this monograph. The conclusion is significant, as it has been shown the high value drivers and traffic agencies give to travel time information in order to

1.6 Conclusions and Further Research

support their decisions. In contrast, the difficulties for traffic managers to fund the information infrastructure and to integrate new technologies as they emerge while retaining sufficient homogeneity in the network has also been shown to be challenging. The conclusions of this book match the travel time information desires without falling in the usual requirement of more and more data.

In this last section of this overview, only overall conclusions of the research are presented. Detailed conclusions can be found in each chapter.

A methodology is proposed in which makes use of the available traffic data to provide accurate travel time information in real time to the drivers entering a highway. Measuring is not enough to achieve this objective and very short term forecasting is necessary. The method uses data obtained from toll ticket data (direct measurement) and from a non-intensive loop detector surveillance system (approximately one loop every 5 km) (indirect measurement), and makes the most of a combined use of the data, using a two level data fusion process. The different accuracy of different data sources, their different temporal alignment and their different spatial coverage allows inferring a short term forecast of travel time, which improves the original travel time estimations from a single data source. The proposed method uses toll ticket and loop detector data, but it is not technologically captive. It can be used with any two sources of travel time data, provided that one of them supplies direct measurements (e.g. the innovative Bluetooth signature matching, or the cell phone tracking) while the other indirect estimations. The results of the data fusion process improve with the accuracy of the single source measurements. Travel time estimation methods from loop detector data are analyzed.

In the process to fuse different travel time estimations, it is found that the current research trend, based in looking for new mathematical speed interpolation methods between point measurements in order to solve the problem of punctual measurements while requiring spatial results (i.e. travel times), it is not adequate if it is blind to traffic dynamics. The present book demonstrates conceptually and with an accurate empirical comparison that travel time estimation methods based on mathematical speed interpolations between measurement points, which do not consider traffic dynamics and the nature of queue evolution, do not contribute in an intrinsically better estimation, independently of the complexity of the interpolation method. Incorporating the traffic dynamics of the theory of kinematic waves, it could improve the performance of these methods. Other technological problems remain, like the frequent failures of loop detectors or their inability to correctly measure over instable stop and go traffic. These issues need to be considered in the future development of the new detectors which will finally beat traditional loops. In addition they will need to be cheap, will need to benefit from a long lifespan, will need to be less intrusive into the pavement, will need to require low maintenance, will need to communicate wireless and will not need a power supply. The MeMS detectors, developed by UC Berkeley engineers, are on the track. This will be the time to incorporate to these detectors more advanced features, like reidentification capabilities. This last issue, it is surely a shared responsibility with the automotive industry, which will need to equip vehicles with an electronic wireless readable tag as an standard equipment.

Travel time estimation from the reidentification of toll tickets in a closed toll highway is also addressed in the monograph. A method is proposed capable of estimating single section travel times (i.e. time required to travel between two consecutive junctions on the main trunk of the highway) and also the exit time at each junction (i.e. the time required to travel along the exit link plus the time required to pay the fee at the toll gate). Combining both estimations it is possible to calculate all the required itinerary travel times, even those with very few observations where direct measurement would be problematic, and avoiding the information delay for real time application. The knowledge of the exit time, allows obtaining the level of service of each toll plaza at every junction, making possible to modify the number of active toll booths in accordance. An extension of the method to open toll highway schemes is currently under research. The key issue here is the optimal location of additional control points (i.e. in addition to main trunk toll plazas), to achieve the most cost—effective surveillance scheme.

The proposed method is only one of the possible applications of the enormously rich database provided by toll ticket data in closed toll highways. A closed toll highway is a privileged infrastructure (in terms of data), where the origin, destination, type of vehicle and entrance/exit times are measured in real time and for all vehicles. This is inconceivable in any other road environment, and will fulfill the desires of the most exigent highway engineer. Tolling is the data reason to exist, and all further use of the data will contribute in net benefits. The potential of the data is huge, for researchers who would like to use the infrastructure as a highway lab to test and evaluate their models, and even for the operators which may apply most of the advances in traffic engineering with privileged data inputs, which are the main drawback of these techniques in most of the applications.

Making use of this privileged situation, all the methods proposed in the monograph have been empirically tested with data obtained in the AP-7 highway. Promising results have been obtained. Now that the conceptual developments have been exposed, it is the turn of highway operators and administrations to put them into practice, so that highway users can benefit of real time travel time estimations with a low-cost scheme. This could also be the seed for turning this test site into a real highway lab and detailed traffic observatory (similar or even better than the Berkeley Highway Lab, a current outstanding example), encouraging further research in information and operations on highway environments.

Other research directions should not be shelved. The highway system efficiency is only one leg of the optimal transportation system. Further research is also deserved in integrated corridor management policies, not only accounting for vehicular traffic, but also for other transportation modes. The optimal supply share of the corridor infrastructures between modes should be a target. Besides, future research should not only consider vehicle to infrastructure communications (as presented in the present monograph), but also exploit the huge potential of the futuristic vehicle to vehicle information systems, which in the next future would, by sure, become a reality.

References

Abdulhai, B., & Tabib, S. M. (2003). Spatio-temporal inductance-pattern recognition for vehicle reidentification. *Transportation Research Part C, 11*(3–4), 223–239.

Barceló, J., Montero, L., & Marquès, L. (2010). Travel time forecasting and dynamic OD estimation in freeways based on Bluetooth traffic monitoring. In *Proceedings of the 89th Annual Meeting of the Transportation Research Board*, Washington D.C.

Berechman, J. (2003). Transportation–Economic aspects of Roman highway development: The case of Via Appia. *Transportation Research Part A, 37*(5), 453–478.

Bloch, I. (1996). Information combination operators for data fusion: Comparative review with Classification. *IEEE Transactions on Systems, Man and Cybernetics–Part A* **26**(1).

Blue, V., List, G., & Embrechts, M. (1994). Neural net freeway travel time estimation. *Proceedings of Intelligent engineering Systems through Artificial Neural Networks, 4*, 11135–11140.

Buisson, C. (2006). Simple traffic model for a simple problem: Sizing travel time measurement devices. *Transportation Research Record: Journal of the Transportation Research Board, 1965*, 210–218.

Cassidy, M. J., & Windover, J. R. (1995). Methodology for assessing dynamics of freeway traffic flow. *Transportation Research Record: Journal of the Transportation Research Board, 1484*, 73–79.

Chen, C., Kwon, J., Skabardonis, A., & Varaiya, P. (2003). Detecting errors and imputing missing data for single loop surveillance systems. *Transportation Research Record: Journal of the Transportation Research Board, 1855*, 160–167.

Chrobok, R., Kaumann, O., Wahle, J., & Schreckenberg, M. (2004). Different methods of traffic forecast based on real data. *European Journal of Operational Research, 155*, 558–568.

Clark, S. D., Dougherty, M. S., & Kirby, H. R. (1993). *The use of neural network and time series models for short term forecasting: A comparative study*. Manchester: Proceedings of PTRC.

Coifman, B., & Ergueta, E. (2003). Improved vehicle reidentification and travel time measurement on congested freeways. *ASCE Journal of Transportation Engineering, 129*(5), 475–483.

Coifman, B., & Krishnamurthya, S. (2007). Vehicle reidentification and travel time measurement across freeway junctions using the existing detector infrastructure. *Transportation Research Part C, 15*(3), 135–153.

Coifman, B., & Cassidy, M. (2002). Vehicle reidentification and travel time measurement on congested freeways. *Transportation Research Part A, 36*(10), 899–917.

Daganzo, C. F. (1983). Derivation of delays based on input-output analysis. *Transportation Research Part A, 17*(5), 341–342.

Daganzo, C. F. (1997). *Fundamentals of transportation and traffic operations*. Elsevier Science Ltd: Pergamon.

Daganzo, C. F. (1995). Requiem for second-order fluid approximations of traffic flow. *Transportation Research Part B, 29*(4), 277–286.

Dailey, D. J. (1993). Travel time estimation using cross-correlation techniques. *Transportation Research Part B, 27*(2), 97–107.

Danech-Pajouh, M. (2003). Les modèles de prévision du dispositif Bison futé et leur évolution. *Recherce Transports Sécurité, 78*, 1–20. In French.

Dia, H. (2001). An objected-oriented neural network approach to short-term traffic forecasting. *European Journal of Operational Research, 131*, 253–261.

Dougherty, M. S., & Cobbett, M. R. (1997). Short-term inter-urban traffic forecasts using neural networks. *International Journal of Forecasting, 13*, 21–31.

Federal Highway Administration. (2010). *UCR—Urban congestion report*. U.S. Department of Transportation.

Hall, D., & McMullen, S. A. H. (2004). *Mathematical techniques in multisensor data fusion*. Boston: Archtech House Publishers.

Harder, K. A., Bloomfield, J. R., Levinson, D. M., & Zhang, L. (2005). *Route preferences and the value of travel-time information for motorists driving real-world routes.* University of Minnesota Department of Civil Engineering.

Herrera, J. C., Work, D. B., Herring, R., Ban, X., Jacobson, Q., & Bayen, A. M. (2010). Evaluation of traffic data obtained via GPS-enabled mobile phones: The mobile century field experiment. *Transportation Research Part C, 18*(4), 568–583.

Highway Research Board. (1950). *Highway capacity manual: Practical applications of research.* National Research Council, Washington D.C.

Hill, J., Szewczyk, R., Woo, A., Hollar, S., Culler, D., & Pister, K. (2000). System architecture directions for networked sensors. *Proceedings of ASPLOS, 2000*, 93–104.

Huang, T. & Russell, S. (1997). Object identification in a Bayesian context. In *Proceedings of the Fifteenth International Joint Conference on Artificial Intelligence (IJCAI-97)*, Nagoya, Japan. Morgan Kaufmann.

Kuhne, R. & Immes, S. (1993). Freeway control systems using section-related traffic variable detection. In *Proceedings of Pacific Rim TransTech Conference* (pp. 56–62), Vol. **1**.

Kwon, T. M. (2006). Blind deconvolution processing of loop inductance signals for vehicle reidentification. In *Proceedings of the 85th Annual Meeting of the Transportation Research Board*.

Lucas, D. E., Mirchandani, P. B., & Verma, N. (2004). Online travel time estimation without vehicle identification. *Transportation Research Record: Journal of the Transportation Research Board, 1867*, 193–201.

MacCarley, A. C. (2001). *Video-based vehicle signature analysis and tracking system phase 2: Algorithm development and preliminary testing.* California PATH Working Paper, UCB-ITS-PWP-2001-10.

Makigami, Y., Newell, G. F., & Rothery, R. (1971). Three-dimensional representations of traffic flow. *Transportation Science, 5*, 302–313.

Muñoz, J. C., & Daganzo, C. F. (2002). The bottleneck mechanism of a freeway diverge. *Transportation Research Part A, 36*(6), 483–505.

Nahum, P. J., & Pokoski, J. L. (1980). NCTR plus sensor fusion equals IFFN. *IEEE Transactions on Aerospace Electronic Systems, 16*(3), 320–337.

Nam, D. H., & Drew, D. R. (1996). Traffic dynamics: Method for estimating freeway travel times in real time from flow measurements. *ASCE Journal of Transportation Engineering, 122*(3), 185–191.

Newell, G. F. (1982). *Applications of queuing theory* (2nd ed.). London: Chapman and Hall.

Newell, G. F. (1993). A simplified theory of kinematic waves in highway traffic, I general theory, II queuing at freeway bottlenecks, III multi-destination flows. *Transportation Research Part B, 27*(1), 281–313.

Nishiuchi, H., Nakamura, K., Bajwa, S., Chung, E., Kuwahara, M. (2006). Evaluation of travel time and OD variation on the Tokyo Metropolitan Expressway using ETC data. *Research into Practice: 22nd ARRB Conference Proceedings Information*, Australian Road Research Board.

NYSI&IS. (1970). Automatic license plate scanning (ALPS) system—Final report. New York State Identification and Intelligence System, Albany, NY.

OECD/JTRC. (2010). Improving reliability on surface transport networks. Paris: OECD Publishing.

Oh, J. S., Jayakrishnan, R., Recker, W. (2003). Section travel time estimation from point detection data. In *Proceedings of the 82nd Transportation Research Board Annual Meeting.* Washington D.C.

Palen, J. (1997). The need for surveillance in intelligent transportation systems. *Intellimotion, 6*(1), 1–3.

Petty, K. F., Bickel, P., Ostland, M., Rice, J., Schoenberg, F., Jiang, J., & Rotov, Y. (1998). Accurate estimation of travel time from single loop detectors. *Transportation Research Part A, 32*(1), 1–17.

References

Robusté, F., Vergara, C., Thorson, L., & Estrada, M. (2003). Nuevas tecnologías en la gestión de autopistas: El peaje y los sistemas inteligentes de transporte. *Revista de economía industrial, 353*, 33–46. In Spanish.

Rosas, D., F. Soriguera & E. Molins. (2008) Construcción de patrones de tiempo de viaje por carretera. In *Proceedings of the VIII Congreso de Ingeniería del Transporte,* La Coruña. In Spanish.

Sherali, H. D., Desai, J., & Rakha, H. (2006). A discrete optimization approach for locating Automatic Vehicle Identification readers for the provision of roadway travel times. *Transportation Research Part B, 40*(10), 857–871.

Soriguera, F., Molins, E., Alberich, E. (2008). Metodología de elaboración de patrones de intensidad de tráfico. In *Proceedings of the VIII Congreso de Ingeniería del Transporte,* La Coruña. In Spanish.

Soriguera, F., Thorson, L., Grau, J. A., & Robusté, F. (2006) Bases de un sistema de medición y previsión del tiempo de viaje por carretera en Cataluña. In *Proceedings of the VII Congreso de Ingeniería del Transporte,* Ciudad Real. In Spanish.

Turner, S. M., Eisele, W. L., Benz, R. J., Holdener, D. J. (1998). *Travel time data collection handbook.* Research Report FHWA-PL-98-035. Federal Highway Administration, Office of Highway Information Management, Washington, D. C.

van Arem, B., van der Vlist, M. J. M., Muste, M. R., & Smulders, S. A. (1997). Travel time estimation in the GERIDIEN project. *International Journal of Forecasting, 13,* 73–85.

Van Iseghem, S. (1999). Forecasting traffic one or two days in advance. *An intermodal approach. Recherce Transports Sécurité, 65,* 79–97.

Wardrop, J. G. (1952). Some theoretical aspects of road traffic research. *Proceedings of the Institute of Civil Engineers, 2,* 325–378.

Chapter 2
Travel Time Definitions

Abstract In this Chapter, travel time definitions are analytically presented. Also, a trajectory reconstruction algorithm necessary in order to navigate between different travel time definitions is proposed. The concepts presented in this chapter are aimed to create a conceptual framework useful in comparing travel times obtained from different methodologies. This should be considered as baseline knowledge when going through the whole book.

2.1 Introduction

All the studies dealing with travel time estimation compare the results of their proposed methods to some ground truth travel time data in order to evaluate the accuracy of the method. In fact, some studies are only devoted to that comparison (Li et al. 2006; Kothuri et al. 2007, 2008). The nature of the ground truth travel time data used in each study is varied. For instance Kothuri et al. (2007) uses data obtained from probe vehicle runs, and Kothuri et al. (2008) adds data from the bus trajectories obtained from a GPS equipped bus fleet. In addition to the probe vehicle runs, Sun et al. (2008) also considers travel time data obtained from video camera vehicle recognition. Li et al. (2006) also uses data obtained from vehicle reidentification at control points, in this case by means of toll tags and number plate matching, while Coifman (2002) uses the length of vehicles to reidentify them from double loop detector measurements. In all cases, ground truth travel time data are obtained by directly measuring travel times, whether tracking the vehicle or identifying it at two successive control points. The ways in which these ground truth travel time data are obtained have some implications in the comparison procedure. Coarse comparisons can lead to the counterintuitive results found in literature because what is being compared are apples and oranges.

In the absence of these directly measured travel time data, the alternative can be simulated data using traffic microsimulators (Cortés et al. 2002; van Lint and van der Zijpp 2003). The same care must be taken with the simulated data, and in addition, it must take into account that the simulation is a simplification of the real traffic dynamics, and may have not been considering all the complexities of real traffic, resulting in predictable evolution of travel times. This leads to an artificial improvement of travel time estimation methods when ground truth data is obtained from simulation (Li et al. 2006).

The present chapter aims to rigorously present travel time definitions in order to fully understand the nature of each type of measurement. The first step is to differentiate between link (or section) travel time in relation to corridor (or itinerary) travel times. A link is the shortest freeway section where travel time can be estimated, while the corridor refers to the target itinerary whose travel time information is useful to the driver. The common practice (and the case in the present monograph) is to define links limited by a pair of detector sites (this represents only some hundreds of meters in metropolitan freeways), while itineraries are defined, for instance, between freeway junctions. Therefore, an itinerary is usually composed of several links.

2.2 Link Travel Time Definitions

Consider the highway link of length "Δx" and the time interval of data aggregation "Δt" shown in Fig. 2.1. In this configuration, the true average travel time over the space-time region "$A = (\Delta x, \Delta t)$" can be expressed as:

$$T_T(A) = \frac{\Delta x}{v(A)} = \Delta x \cdot \sum_{i=1}^{n} \frac{d_i}{T_i} \qquad (2.1)$$

where:
"$v(A)$" is the generalized average speed definition in the region "A", first proposed by Edie (1965).
"d_i" is the distance traveled by the ith vehicle in the region "A".
"T_i" is the time spent by the ith vehicle in the region "A".

Note that two control points at "x_{j-1}" and "x_j" where vehicles are identified are not enough to obtain this true average travel time, as the position of the vehicles travelling within the section at time instants "$p - \Delta t$" and "p" could not be obtained (i.e. points where vehicles cross the time borders of region "A" in the time-space diagram —see Fig. 2.1a). It is possible that the only way to directly measure this travel time is by continuously tracking all the vehicles (or a representative sample of them).

The true average travel time "$T_T(A)$" should not be confused with the arrival based average travel time, "$T_A(A)$", defined as the average travel time in the trip along the whole link "j" of those vehicles that reach "x_j" in the time period "p" (see Fig. 2.1b). This type of ground-truth travel time is obtained from all the direct measurements based on the reidentification of vehicles (number plates, toll tags, bluetooth devices, electromagnetic signatures, platoons, cumulative counts …).

2.2 Link Travel Time Definitions

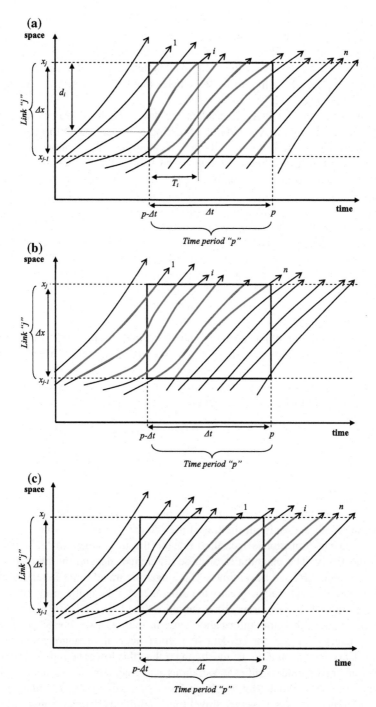

Fig. 2.1 Link travel time definitions in a trajectories diagram. **a** True average travel time. **b** Arrival based average travel time. **c** Departure based average travel time. *Note* In *red*, parts of the vehicles' trajectories considered in the average travel time definitions

As this is the most common directly measured type of travel time, it is sometimes named MTT (measured travel times).

Following the same logic, a third average travel time can be defined. The departure based average travel time, "$T_D(A)$", is defined as the average travel time on a trip along the whole link of those vehicles that depart from "x_{j-1}" in the time period "p" (see Fig. 2.1c).

On the one hand, "$T_A(A)$" considers the last completed trajectories on the highway link, and this may involve considering relatively old information of the traffic conditions on the first part of the link (some of the information was obtained more than one travel time before). On the other hand, "$T_T(A)$" uses the most recent information obtained in the whole link (sometimes these types of travel times are named ITT—instantaneous travel time). However, it is possible that any vehicle has followed a trajectory from which this true average travel time results. Finally, "$T_D(A)$" needs future information in relation to the instant of calculation. Therefore it is not possible to compute "$T_D(A)$" in real time operation. However, there is no problem in obtaining this future estimation in an off-line basis, when a complete database is available, including future information in relation to the instant of calculation. Note that "$T_D(A)$" would be approximately equal to "$T_A(A')$" where "A'" corresponds to the space-time region "A" moved forward one travel time unit in the time axis.

It is also possible, but not so easy, to obtain "$T_A(A)$" and "$T_D(A)$" from "$T_T(A)$". It is only necessary to compute the position of a virtual vehicle within the link as a function of time and considering the average speeds resulting from "$T_T(A)$" at different time intervals. This process, known as trajectory reconstruction, is detailed in Sect. 2.4. Note, that in order to obtain "$T_D(A)$", future "$T_T(A)$" will be needed.

2.2.1 When Travel Time Definition Makes a Difference

The differences between these average travel time definitions lie in the vehicle trajectories considered in the average calculation, "$T_T(A)$" being the only definition that considers all and only all the trajectories contained in "A", while "$T_A(A)$" or "$T_D(A)$" consider trajectories measured outside the time edges of "A", before or after respectively. The magnitudes of these differences depend on the relative difference between "Δt" and travel times. The longer the travel times in relation to the updating time interval are, the greater the difference will be in the group of vehicles considered in each average travel time definition (see Fig. 2.2). "Δt" is a parameter to be set for the travel time information system, with a lower bound equal to the updating interval of the source data (e.g. time interval of aggregation of loop detector data). "Δt" should not be much longer than this lower bound in order not to smooth out travel time significant variations and maintain an adequate updating frequency (i.e. "Δt" should not go above 5 min). Therefore, as "Δt" must be kept small, differences between average travel time definitions depend mainly on travel

2.2 Link Travel Time Definitions

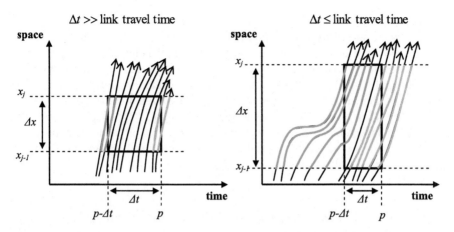

Fig. 2.2 Different trajectories considered in the link travel time definitions. *Note* In *blue*, trajectories considered in the arrival based average travel times but not in the departure based average travel times. In *orange*, the opposite situation, both types of trajectories are only partially considered in the true average travel time. In *black*, shared trajectories

times, which in turn, depend on the length of the highway link, "Δx", and on the traffic conditions.

In situations where link travel times are significantly longer than "Δt", as would happen in the case of long highway links or in the case of congested traffic conditions, the trajectories considered in several average travel time calculations will belong to different groups of vehicles (see Fig. 2.2, right). The case may even arise where none of the vehicle trajectories are shared between different definitions. This would not have any effect on the average travel time in the case of stationary traffic, as the trajectories of the different groups of vehicles would be very similar. However, if a traffic transition occurs in the space-time regions considered in one definition but not in the others, this could result in significant differences between computed average travel times. This is the situation when the definition of average travel time plays an important role. On the contrary, in situations where link travel times are significantly shorter than "Δt" (i.e. short highway links due to high surveillance density and free flowing traffic conditions), the vehicle trajectories considered in one definition but not in the others would be very limited in relation to the total amount of shared trajectories (see Fig. 2.2, left). Therefore, the probability and the relative weight of traffic transitions in this reduced space-time region is very low. This results in differences among definitions as being almost negligible in this case. From this discussion it is concluded than when real-time freeway travel time information is most valuable (i.e. congested and evolving traffic conditions) the different definitions of average travel time play an important role which must be considered.

Also note that it is rather difficult in practice to obtain "$T_T(A)$". However, to obtain "$T_A(A)$", only two vehicle identification points are necessary. As a result of this, there seems to be an interesting possibility of obtaining an approximation to "$T_T(A)$" by using the measurements that configure "$T_A(A)$". This approximation is as simple as only considering the trajectories which have arrived at the downstream control point during "p" time period (i.e. they belong to "$T_A(A)$" group of trajectories) and have departed from the upstream control point also during "p" (i.e. trajectories fully contained in "A" or equally, the shared trajectories between "$T_D(A)$" and "$T_A(A)$", see Fig. 2.2). This approximation would be better as the number of shared trajectories increase. The process would converge to a perfect estimation when all trajectories are shared (i.e. "$T_T(A)$", "$T_A(A)$" and "$T_D(A)$" are equal). On the contrary, in some situations the approximation cannot be applied due to the inexistence of shared trajectories. This would result in the possibility of "$T_A(A)$" being a bad approximation to "$T_T(A)$".

2.2.2 Which Information Is Actually Desired from Real Time Systems?

In a real-time information dissemination scheme, "$T_T(A)$" and "$T_A(A)$" would be available for drivers entering the section at time period "$p + 1$". However, neither the true average travel time, nor the arrival based average travel time at time interval "p" is the information that these drivers wish to obtain. They want to know their expected travel time, and therefore a departure based travel time at time interval "$p + 1$" (sometimes this travel time is known as a PTT—predicted travel time). Therefore, the desired forecasting capabilities of measured travel time must not only span a time horizon equal to the travel time (i.e. in order to obtain the departure based average travel time at time period "p"), but an extended horizon equal to the travel time plus "Δt". This leads to the apparent paradox that while for a longer "Δt"s the true average travel time measurement is a better approach to the departure based average travel time at time period "p" (because more trajectories will be shared), the error made with the naïve assumption of considering "$T_T(A)$" as a proxy for the departure based travel time at time interval "$p + 1$" (the implicit assumption here is that traffic conditions on the corridor remain constant from the measuring instant until the forecasting horizon) usually increases as "Δt" does (because of the extension of the forecasting horizon). Therefore, in some contexts (i.e. transitions, when the implicit assumption does not hold) the averaging of traffic conditions within long "Δt"s could lead to huge variations between adjacent time intervals. This is another reason for "Δt" being kept short.

2.3 Corridor Travel Time

The presented link average travel time definitions are also valid in a corridor context. The main difference in this case is that while the "Δt" remains the same as in the link basis, the increased length of the corridor in relation to the several links from which it is constituted of individual length equal to "Δx_j" results in larger

2.3 Corridor Travel Time

travel times. This implies "$T_T(C)$", "$T_A(C)$" and "$T_D(C)$", where "C" stands for a corridor space—time region "$C = \left(\sum_{j \in corridor} \Delta x_j, \Delta t\right)$" being significantly different in non stationary traffic conditions.

Another issue to consider is how the corridor travel times could be obtained from composing link travel times. In can be easily deduced that corridor "$T_T(C)$" is obtained by simply adding up the links "$T_T(A)$" from the time period of calculation. On the contrary, corridor "$T_A(C)$" and corridor "$T_D(C)$" are not obtained from this simple addition, as it is needed to consider the vehicle trajectory in space and time.

As a conclusion to this definitions section, take into account that the common practice in the real time implementation of travel time systems based on speed point measurements is to estimate link "$T_T(A)$" by means of the available loop detectors, which are added up to obtain the corridor travel time, "$T_T(C)$", to be disseminated in real time. These true average corridor travel times are considered to be the best measurable estimation for the desired departure based average travel time at time interval "$p + 1$", if one wants to avoid the uncertainties of forecasting, and assumes traffic conditions will remain constant. In particular, better than the "delayed" information from arrival based average travel times is "$T_A(C)$" (which could be directly measured in the corridor), provided that "$T_T(A)$" estimations of speed are accurate. Otherwise, this assertion could not be true. Also note that differences between true and predicted travel times depend on the corridor length and on the aggregation period "Δt", which constitute the horizon of the prediction. Therefore, in order to keep differences low and improve the accuracy of the "forecast", an advisable dissemination strategy is to keep corridor lengths as short as possible, while maintaining the interest of the driver on the disseminated information, and frequent updating, so that the time horizon of the prediction is as short as possible.

In the case of off-line travel time assessment, there is no need of trying to infer future travel times, for example by considering the latest information on the corridor as "$T_T(C)$" does. However, it is advisable to assess the real travel time that drivers actually experimented; this means reconstructing their trajectories in order to obtain "$T_A(C)$" from original "$T_T(C)$". These measurements are different in nature, and although they may be pretty similar in a link context, they will be significantly different on a corridor basis and non stationary traffic conditions. The results would be analogous to those obtained from direct measurement from an AVI device, provided that the original true link average travel times were accurate. Therefore, the same process of time and space alignment must be undertaken in a case of comparisons between "$T_T(C)$" obtained from loop detectors and "$T_A(C)$" obtained from AVI direct measurements. This process is described in detail in the next section.

2.4 Trajectory Reconstruction Process

This section aims to present the simple process necessary in order to reconstruct a vehicle trajectory from a speeds field in a discretized space-time plane. In other words, if space mean speeds are available within each link as a function of "x"

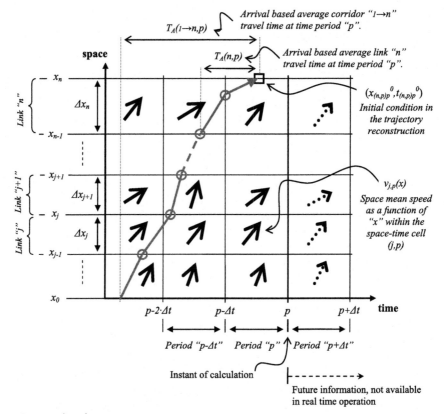

Fig. 2.3 Speeds field in a space-time discretization

(i.e. the position of the virtual vehicle within the link), and this function "$v(x)$", it is assumed that it will remain constant within each time interval "Δt" (see Fig. 2.3). It is possible to reconstruct the trajectories that would result from the arrival based or departure based average link and corridor travel times. It is also possible to obtain the link and corridor average true travel time.

Therefore, the following process is necessary in order to convert true (or sometimes called "instant") travel times into trajectory based travel times. This process is analogous for the arrival based (backward reconstruction) or departure based (forward reconstruction) travel time averages. Taking into account that departure based travel times require future true information, only backwards reconstruction will be described in detail, but the analogous process can be easily derived (van Lint and van der Zijpp 2003).

2.4 Trajectory Reconstruction Process

As van Lint and van der Zijpp (2003) describe, to reconstruct the trajectory of a virtual vehicle [i.e. to obtain the function "$x(t)$"] within a cell (j, q) of the speeds field it is only necessary to solve the following differential equation:

$$\frac{\partial x}{\partial t} = v_{j,q}(x) \qquad (2.2)$$

Given "$v_{j,q}(x)$", which do not depend on time within a particular cell, and an initial condition "$x(t_{j,q}^0) = x_{j,q}^0$", which in this backward trajectory reconstruction corresponds to the cell exit point of the trajectory. The obtained solution will be valid for that particular cell. In order to obtain the whole trajectory along the link or the corridor to compute the arrival based travel time in time interval "p", it is only necessary to apply the aforementioned process iteratively from the last cell which crosses the trajectory, (n, p) (see Fig. 2.3), until the start of the corridor is reached. At each step the "$v_{j,q}(x)$" (which is a function of the cell and of the time interval) and the initial condition must be updated. Note that the initial condition for subsequent cells corresponds to the entrance point in the space-time diagram of the trajectory to the previously calculated cell (see Fig. 2.3). Then, the only initial condition needed to be set is the first one, corresponding to the time instant of calculation of the average corridor travel time. It seems adequate to consider this

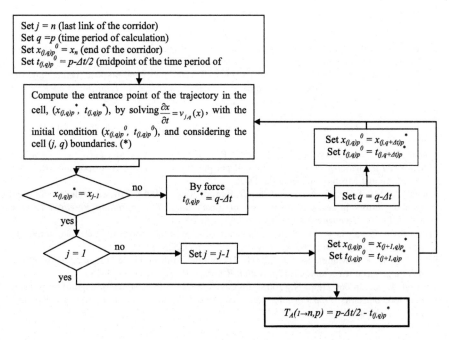

Fig. 2.4 Trajectory reconstruction flow chart. (*Asterisk*) This calculation is explained in detail in the next section for each type of function defining $v(x)$

first initial condition, when computing the arrival based average corridor "$1 \to n$" travel time at time period "p", as the midpoint of the time interval:

$$x(p - \Delta t/2) = x_n \qquad (2.3)$$

The remaining initial conditions, as each cell is confined by space and time bounds, will be defined by the instant the virtual vehicle crosses a link border, or the position within the link where the vehicle undergoes a change of time period.

The whole process for time interval "p" is detailed in a flowchart in Fig. 2.4.

The computation of the true average link travel time for link "j" and time period "p", "$T_T(j, p)$", is simpler as it is only needed to solve Eq. 2.2 without considering the time boundary of the cell. On each link, the initial condition could be "$x(p) = x_j$". Once the equation has been solved and the trajectory function "$x(t)$" is obtained, "$T_T(j, p)$" is calculated by imposing "$x(p - T_T(j,p)) = x_{j-1}$".

Finally, the true average corridor "$1 \to n$" travel time for the time period "p" is obtained as:

$$T_T(1 \to n, p) = \sum_{j=1}^{n} T_T(j, p) \qquad (2.4)$$

References

Coifman, B. (2002). Estimating travel times and vehicle trajectories on freeways using dual loop detectors. *Transportation Research Part A, 36*(4), 351–364.

Cortés, C. E., Lavanya, R., Oh, J., & Jayakrishnan, R. (2002). General-purpose methodology for estimating link travel time with multiple-point detection of traffic. *Transportation Research Record: Journal of the Transportation Research Board, 1802,* 181–189.

Edie, L. C. (1965). Discussion of traffic stream measurements and definitions. In J. Almond (Ed.) *Proceedings of 2nd International Symposium on the Theory of Traffic Flow,* OECD, Paris (pp. 139–154).

Kothuri, S. M., Tufte, K. A., Ahn, S., & Bertini, R. L. (2007). *Using archived ITS data to generate improved freeway travel time estimates.* Paper presented in the 86th Transportation Research Board Annual Meeting. Washington D.C.

Kothuri, S. M., Tufte, K. A., Fayed, E., & Bertini, R. L. (2008). Toward understanding and reducing errors in real-time estimation of travel times. *Transportation Research Record: Journal of the Transportation Research Board, 2049,* 21–28.

Li, R., Rose, G., & Sarvi, M. (2006). Evaluation of speed-based travel time estimation models. *ASCE Journal of Transportation Engineering, 132*(7), 540–547.

Sun, L., Yang, J., & Mahmassani, H. (2008). Travel time estimation based on piecewise truncated quadratic speed trajectory. *Transportation Research Part A, 42*(1), 173–186.

van Lint, J., & van der Zijpp, N. (2003). Improving a travel-time estimation algorithm by using dual loop detectors. *Transportation Research Record: Journal of the Transportation Research Board 1855,* 41–48.

Chapter 3
Accuracy of Travel Time Estimation Methods Based on Punctual Speed Interpolations

Abstract The accuracy of real-time travel time information disseminated on metropolitan freeways is one of the key issues in the development of advanced traveler information systems. Although very accurate estimations could be obtained if suitable and intensive monitoring systems were available, travel time estimations must usually rely on data obtained from the preexisting surveillance equipment installed on freeways: loop detectors. Travel time estimation from loop measurements has attracted extensive research in the last decade, resulting in numerous methodologies. Among these, the ones that rely on spot speed measurements at detector sites in order to obtain the travel time estimation on the target stretch are the most intuitive. The key issue concerning these methods is the spatial generalization of point measurements over a freeway link. Multiple approaches can be found in the literature, ranging from the simplest, and mostly implemented in practice, constant speed approach, to recent and more complex mathematical interpolations. The present chapter shows that all speed interpolation methods that omit traffic dynamics and queue evolution do not contribute to better travel time estimations. All methods are inaccurate in congested and transition conditions, and the claimed relative benefits using various speed interpolation methods result from context specific experiments. Therefore, these methods should be used carefully, and not taken as perfect. Lacking a better approach, it is recommended to avoid overcomplicated mathematical interpolations and focus the efforts on intelligent smoothing of the noisy loop detector data, reducing the fluctuations of short time interval aggregations while maintaining the immediacy of the measurements.

Keywords Travel time estimation · Loop detector data · Speed trajectory interpolation

3.1 Introduction and Background

Information on the expected travel time along a congested freeway corridor is perhaps the most valuable traffic information for commuters in order for them to improve the quality and efficiency of their trips (Palen 1997). Pre-trip information may allow drivers the selection of time and route or even a mode shift. On-trip information is valuable for rerouting or deciding to accept park&ride options. In both situations, travel time information contributes to congestion mitigation. Even in the case where no travel time improvement was possible, travel time information would still improve the quality of the journey by reducing the uncertainty and consequently the stress of the driver. It must be highlighted that the accuracy of the disseminated information is crucial, as providing inaccurate travel time estimations can be detrimental.

Not only drivers benefit from travel time information, but also highway administrations, as travel time is an essential quantitative variable for evaluating the performance of transportation networks, or the operational efficiency of traffic management strategies, and can also be used as a robust and deterministic indicator of an incident. Travel time will become a basic input for the new real-time Advanced Traffic Management Systems (ATMS).

The abbreviation ATIS (Advanced Traveler Information Systems), groups all the technological elements necessary in order to develop a travel time information system, from the measurement of the source data to the dissemination of final information. While the deployment of technological equipment involved in the dissemination of variable traffic information moves toward a positive end (e.g. on-board traffic information devices are best-sellers, Variable Message Signs—VMS—are being widely installed on metropolitan freeways, traffic information web sites are becoming more popular, ...) (OECD/JTRC 2010), the measurement of the source travel time data is more problematic. In order to directly measure the travel time of vehicles on a freeway section, area wide monitoring is required. This means that it is necessary to record the position of the vehicle every few seconds (i.e. vehicle tracking), which could be achieved using traditional probe cars, or in its new concept by tracking GPS equipped vehicle fleets. In order to obtain a continuous flow of travel time measurements, a high percentage of equipped vehicles is necessary. Presently, this limits the practical application of this measurement technology to particular and delimited travel time experiments (Turner et al. 1998), although great hope exists for vehicle tracking once some type of sensor becomes extremely popular, for example GPS equipped cell phones (Claudel et al. 2009).

The alternative for the direct measurement of freeway travel times is not measuring the detailed trajectory of the vehicle, but only identifying the times the vehicle enters and exits the target section. Although this simplification has some implications on the nature of the measurements (see Chap. 2 for details), it allows a direct measurement of travel times by only identifying the vehicle at two control points. Classical AVI (Automated Vehicle Identification) technologies, such as video recognition of license plates or automatic reading of toll tags (in case of a

3.1 Introduction and Background

turnpike highway equipped with an Electronic Collection System—ETC—device) (see Chap. 5) are already being used. Furthermore, promising innovative schemes like the identification of the bluetooth signature of a particular vehicle must also be considered.

In spite of this technological feasibility and accuracy of directly measuring travel times using area wide surveillance equipment, most of the Traffic Management Centers (TMCs) around the world currently rely on traditional inductive loop detectors to monitor traffic. TMCs have faced up to the new traffic operations data requirements by increasing the density of detector sites on most metropolitan freeways, up to 1 detector every 0.5 km. Therefore, any travel time estimation system hoping to be implemented on a large scale must rely on loop detector point measurements. This situation has not gone unnoticed by the transportation research community and has captured a large research effort during the last decades, as is reflected by the vast number of references that can be found in literature in relation to travel time estimation from loop point measurements.

Conceptually, two research directions could be distinguished. The first approach consists of using loop technology to identify particular vehicles, or groups of vehicles, at consecutive detector sites. Once the vehicles have been reidentified, direct travel time measurements can be obtained. The differences among several methods that fall under this category are in regard to the identification procedure. In Coifman and Cassidy (2002), Coifman and Erguera (2003) and Coifman and Krishnamurthya (2007), the length of vehicles is used to reidentify previously measured patterns. Lucas et al. (2004) reidentifies the platoon based structure of traffic, while Abdulhai and Tabib (2003) uses the particular electromagnetic signature that a vehicle produces when travelling over a loop detector. In practice, none of these methods can be applied using the standard time averaged data that is nowadays sent to the TMC from the detector roadside controller. In Dailey (1993) a statistical method is proposed to match the vehicle count fluctuations around the mean at adjacent detectors. The problem in this case (like in all platoon based identification methods) is that the platoon structure of traffic, which causes the strong correlation in vehicle count fluctuations, is lost in dense traffic or in the case of in-between junctions. Therefore, the aforementioned method is only valid for light free flowing traffic when travel time information is less valuable, as it is already known in advance. In Petty et al. (1998) this drawback is confronted by proposing to search for correlations in a wider and dynamic time window, and obtaining a probability density function of travel times. The authors claim good estimates, even for congested conditions, provided that data time aggregations are on the order of one second. Although they reported promising results, this last condition is not realistic in common present day practices. The last methods which could be grouped in this category are those that use relative differences in cumulative counts at consecutive detectors to estimate travel times (Nam and Drew 1996; Oh et al. 2003). Note that, in fact, this represents a vehicle reidentification under the first in first out assumption. The loop detector count drift is a major drawback in these methods.

The second, but most intuitive and simple category of methods to obtain travel time estimates from loop point measurements, uses the spot speed measurement at the detector site to generalize the speed over the target section and obtain the travel times. As a result of this apparent simplicity and straightforward application using common aggregations of loop detector data, spot speed based travel time estimation methods have been implemented worldwide by most traffic agencies. Problems with these types of methods arise mainly due to two factors: accuracy of spot speed measurements and spatial generalization of point measurements. The lack of accuracy of single loop speed estimations, usually assuming constant vehicle length, is widely accepted and has been vastly analyzed for a long time. Several methods have been proposed to enhance single loop performance in terms of speed measurements (Petty et al. 1998; Mikhalkin et al. 1972; Pushkar et al. 1994; Dailey 1999; Wang and Nihan 2000, 2003; Coifman 2001; Hellinga 2002; Coifman et al. 2003; Lin et al. 2004), however, to solve this problem most highway administrations have chosen to install double loop detectors capable of an accurate speed measurement, at least on their metropolitan freeways where intensive traffic monitoring is crucial.

The spatial generalization of point measurements, necessary in this type of travel time estimation methods, is the second factor that introduces error. The common practice is to assign a particular loop detector to a freeway portion, and assume that the speed remains constant during the whole section and during the whole time aggregation period. Even in the case of a unique stationary traffic state prevailing for the whole freeway section, travel time estimates would be flawed due to the fact that loop detector controllers usually compute and send to the TMC, time mean speeds (e.g. this is the case for the Spanish standards in loop detector data treatment), while the variable that relates average travel time with section length is the space mean speed. As local time mean speed structurally overestimates the space mean because faster observations are overrepresented, average travel times computed in this hypothetical stationary situation will be slightly underestimated (approximately 2 % on average) (Soriguera and Robusté 2011; Li et al. 2006). This drawback could easily be solved by computing the space mean speed at the detector site (i.e. the harmonic mean of individual speeds) instead of the time mean speed (i.e. the arithmetic mean). A more problematic situation, which can be seen as an extreme of this last shortcoming, arises in the case of congested unstable behavior of traffic, when travel time estimation errors using point speed measurements and short aggregation periods can reach 30 % (Rakha and Zhang 2005) even though traffic conditions can be in average the same on the whole freeway link. This is due to the fact that, at the detector site and for the short updating time intervals (<5 min), the stop&go instabilities can result in great errors in measuring a representative speed average for the whole link, because it is possible that the detector only measures average speed over one of the traffic instabilities. Smoothing data over longer time periods or wider measurement regions in space would average together many unstable, non-stationary traffic states, to hopefully converge to an unbiased global average traffic state. In addition in these stop&go situations, the measured average speeds only reflect the "go" part of the traffic and do not account

3.1 Introduction and Background

for the time vehicles are completely stopped. Generally, the "stop" periods are small compared to the travel times and therefore this effect has not a significant contribution. However, in case of very congested traffic states, this last assumption cannot be accepted, and travel time will usually be underestimated, but not always, as it is also possible that the detector is measuring a very low speed instability. These flaws resulting from unstable non-stationary traffic states are not solved by computing the point estimation of space mean speed, as the spatial behavior of traffic instabilities wouldn't be captured either. Therefore, nothing can be said about the effects of using time mean speed instead of space mean speed in these situations. The paradox is that by using the wrong mean speed, improved travel time estimates could happen. However, this has to be considered as a positive accident, and not as a constant rule.

Being aware of these limitations, one must realize that considering common loop detector spacing (i.e. 0.5 km at best), one can see that they are almost negligible in relation to the errors that would arise in travel time estimation in the case of a dramatic traffic state transition on the freeway section (e.g. change from free flowing traffic to queued traffic within the section). Only the stop&go drawback may imply an exception to this last assertion. Take into account that when the spatial stationarity condition is broken, the point measurements assumed for the whole section would be totally unrealistic, and travel time errors can be huge. These will be largest when most of the segment is queued and the detector is unqueued (or vice versa).

This evidence leads to the obvious and widely demonstrated fact (Li et al. 2006; Kothuri et al. 2007, 2008) that travel time estimation methods based on spot speed measurements perform well in free flow conditions (this implies that there isn't any change of a traffic state within the freeway section), while the accuracy of the estimation in congested or transition conditions are dubious (there exists the possibility of a traffic state change within the freeway section and stop&go instabilities). It is also evident that the magnitude of these errors depends highly on the length of the link. For long links (i.e. low surveillance equipment density or increased detector spacing due to the temporary malfunctioning of a particular detector, a situation that arises too often) errors could be enormous, as the erroneous speed spatial generalization would be considered for this long highway section. Moreover, there is more probability of state transition within the link. Therefore, as the length of the links is defined by the surveillance density, there is a clear relation between the distance between loop detectors and the travel time estimation errors.

In order to solve this problem, several authors (Cortés et al. 2002; van Lint and van der Zijpp 2003; Sun et al. 2008; Coifman 2002; Treiber and Helbing 2002) propose new speed interpolation models, different than the constant assumption, to better describe the spatial speed variations between point measurements, especially in congested conditions. Apart from a pair of remarkable exceptions and possible alternatives in (Coifman 2002; Treiber and Helbing 2002), where classical continuum traffic flow theory is used to generalize speed point measurements over the freeway section, the other new models, which are described in detail in Sect. 3.2, are tending to be mathematical exercises of interpolation, blind to traffic stream dynamics. These methods basically smooth the constant interpolation over time or space, but do not address the fact that any feature (i.e. end of a queue location) finer than the detector spacing will go unobserved.

The present chapter aims to demonstrate that there is no reason to expect that a speed interpolation method which does not consider traffic dynamics and queue evolution, performs better in freeway sections that are partially congested. All these methods are inaccurate in congested and transition traffic states and the claimed benefits usually result from context and site specific validations, which sometimes can lead to counterintuitive conclusions (Li et al. 2006). Lacking a better approach (note that the methods proposed in Coifman (2002), Treiber and Helbing (2002) do not contribute in a better approach in partially congested sections, and therefore the problem remains unsolved), the simplest interpolations are recommended and an "intelligent" smoothing process is proposed in order to smooth the typical speed fluctuations of vehicle mean speed over short time intervals and traffic instabilities, while preserving the immediacy characteristic of loop measurements. Surprisingly, not much research effort has been focused on this last issue.

The remainder of this chapter is organized into several sections. Section 3.2 is devoted to the review of the proposed speed interpolation methods, followed by Sect. 3.3 where empirical traffic data from loop detectors and directly measured AVI travel time data are presented, and Sect. 3.4 evaluating the accuracy of the methods presented with the available data. Finally, some conclusions and directions for further research are outlined.

3.2 Methods of Link Travel Time Estimation from Point Speed Measurements

In long interurban trips composed of many links, where only a few suffer from congestion, the extremely inaccurate link travel time estimation in congested or transition conditions could have little effect as traveler does not care about one link's travel time, but on the aggregate travel time along the links that configure his trip. However, in the shorter commute trips across congested metropolitan freeways, where travel time information is more valuable and accuracy crucial, the situation is the other way around. It follows that the key issue in order to accurately estimate corridor travel times in these conditions (both, arrival based average travel times useful for off-line assessment or true average travel time for real time dissemination of information) is an accurate estimation of the true link average travel time within a time period, as true link average travel time is the building block for all other travel time definitions. Therefore the link level is adequate for the analysis.

It has been stated that the main problem in this link travel time estimation from point speed measurements is the lack of knowledge regarding the speed evolution in space, "$v(x)$", between measurement spots. This results in these methods being highly inaccurate when there is a traffic state transition within the link. The magnitude of these errors is directly related to the length of the link, which is inversely proportional to the loop detector density.

The present section is devoted to presenting several proposals for the estimation of "$v(x)$" between detector sites which can be found in the literature or in practical implementations. Explicit formulations to calculate $(x_{j,q}^*, t_{j,q}^*)$, the entrance/exit points of the trajectory in a space-time cell (j, q), will also be derived for each method.

It is interesting to note that some authors (Cortés et al. 2002; Sun et al. 2008) try to estimate "$v(t)$" between detector sites, instead of "$v(x)$". The claimed reason for such an approach is that although assuming continuous and smooth functions represent speed in time and space, the necessary trajectory reconstruction process when using "$v(x)$" results in speed discontinuities at time interval changes. This would not happen in the "$v(t)$" approach because the obtained average link travel time estimation would be directly arrival based, with the related drawbacks in real time estimation. Therefore, the trajectory reconstruction process is not necessary in a link context, but in a corridor basis. In addition, there is no reason to suspect that "$v(x)$" or "$v(t)$" should be continuous and smooth. For instance, a sharp change in speed when a vehicle encounters a queue on a freeway can be seen as a discontinuity in this function between approximate constant speeds. These discontinuities are more intuitive in relation to "x" as they happen on freeway spots where sudden traffic state changes arise. The artificial speed discontinuities every "Δt" within the reconstruction process are an inherent consequence of the discretization of time domain and would be small in the case of frequent updates and accurate estimations of "$v(x)$". Using "$v(t)$" eliminates this drawback indeed, but some complexity is added as the "distance" between interpolation points is not constant, but is rather the precise average link travel time. As the speed measurements at detector spots are not continuous in time (as a consequence of the discrete time domain), there is no guarantee that the iterative process required to obtain "$v(t)$" converges.

3.2.1 Constant Interpolation Between Detectors

The simplest approach for the space generalization of speed between measurement points is the constant speed assumption. For its simplicity, this approach is widely used around the world (van Lint and van der Zijpp 2003; Sun et al. 2008; Kothuri et al. 2008). Several variants exist in relation to which speed measurement is selected to represent the whole section (see Fig. 3.1). For instance, "$V_{j-1,q}$", the upstream speed measurement on the link at "x_{j-1}" and time interval "q" could be considered to represent "$v_{j,q}(x)$" in the whole link "j". Some case specific applications in particular links may suggest that this assumption is acceptable, but in a systematic application there is no objective reason for that. Therefore, on the same

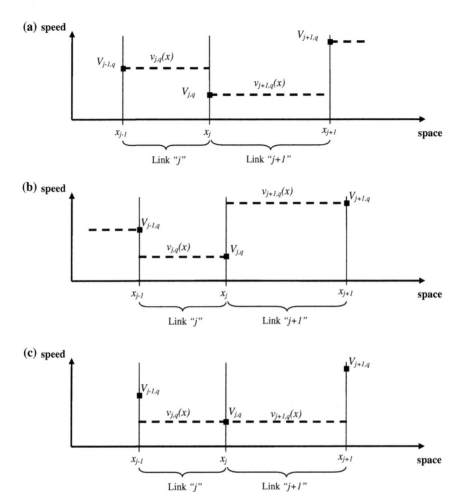

Fig. 3.1 Constant speed trajectory spatial generalization. **a** Upstream, **b** downstream, **c** conservative, **d** optimistic, **e** weighted average

basis, the downstream measurement at "x_j" could be equally valid. Another approach could be to adopt a conservative strategy and assign the whole link the lowest of the speed measurements at "x_{j-1}" and "x_j", as in the ATIS in San Antonio, Texas (Fariello 2002). On the contrary, there could also be an optimistic approach in considering the largest of the measured speeds. Finally, one may also want to adopt an in-between solution (Cortés et al. 2002), and select a weighted average speed "$v_{j,q}(x) = \alpha V_{j-1,q} + (1 - \alpha) \cdot V_{j,q}$" where "$\alpha \in (0, 1)$".

With any of these approaches, the solution to Eq. 2.2 which defines the virtual vehicle trajectory within a space-time cell, (j, q), is expressed as:

3.2 Methods of Link Travel Time Estimation from Point Speed Measurements

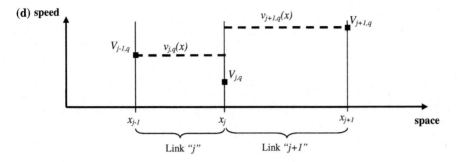

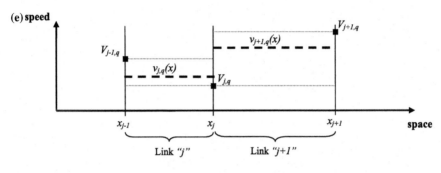

$v_{j,q}(x) = \alpha V_{j-1,q} + (1-\alpha) \cdot V_{j,q}$ with $\alpha \in (0,1)$

Fig. 3.1 (continued)

$$x_{j,q}(t) = x^0_{j,q} + \vartheta_{j,q} \cdot (t - t^0_{j,q}) \qquad (3.1)$$

where $(x^0_{j,q}, t^0_{j,q})$ is the cell exit point of the trajectory and "$\vartheta_{j,q}$" is the selected cell constant speed. From Eq. 3.1, the trajectory entrance point to the cell is:

$$\{x^*_{j,q}, t^*_{j,q}\} = \begin{cases} \left\{x_{j-1}, \frac{x_{j-1}-x^0_{j,q}}{\vartheta_{j,q}} + t^0_{j,q}\right\} & \text{if } x^0_{j,q} + \vartheta_{j,q} \cdot \left((q-\Delta t) - t^0_{j,q}\right) < x_{j-1} \\ \left\{x^0_{j,q} + \vartheta_{j,q} \cdot \left((q-\Delta t) - t^0_{j,q}\right), q - \Delta t\right\} & \text{otherwise} \end{cases}$$

(3.2)

3.2.2 Piecewise Constant Interpolation Between Detectors

A simple modification of the constant speed assumption is the piecewise constant assumption between measurement points. The only difference is that the speed discontinuity is assumed to take place inside the link, and not at detector points. Therefore, the piecewise constant interpolation just redefines where a link begins and ends relatively to the detector. The details of the specific errors should change,

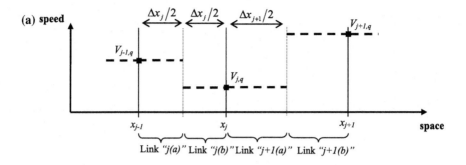

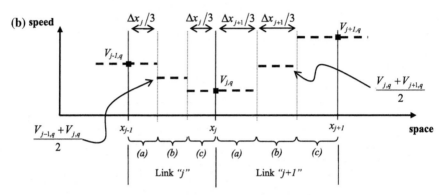

Fig. 3.2 Piecewise constant speed trajectory spatial generalization. **a** Midpoint algorithm. **b** Thirds method

but the net magnitude should not be much different, unless the speed discontinuity location is selected taking into account traffic dynamics within the link. Two common piecewise constant methods are the midpoint algorithm, widely used around the world (Li et al. 2006), and the thirds method (see Fig. 3.2) used by the Minnesota DOT in the Twin Cities metropolitan area (Kwon 2004). In this context, Eqs. 3.1 and 3.2 remain valid, provided that each space-time cell is divided to maintain the constant speed assumption within the cell (see Fig. 3.2).

In some cases the piecewise constant speed trajectory within the link is simplified to a weighted average constant speed interpolation where the weighting factors are each a relative share of the link between speed measurements. Note that the resulting vehicle trajectory, "$x(t)$", reconstructed over the link or the corridor, is different, and therefore trajectory based travel times will be different depending on the method. However, true or instant travel times, which do not depend on the vehicle trajectory, would be the same, provided that the average speed is a harmonic weighed average. The following equation should be applied in order to put the case on a level with the one shown in Fig. 3.1e:

3.2 Methods of Link Travel Time Estimation from Point Speed Measurements

$$T_{T(j,q)} = \frac{\alpha \cdot \Delta x_j}{V_{j-1,q}} + \frac{(1-\alpha) \cdot \Delta x_j}{V_{j,q}} = \frac{\Delta x_j}{\frac{1}{\alpha \cdot \frac{1}{V_{j-1,q}} + (1-\alpha) \cdot \frac{1}{V_{j,q}}}} = \frac{\Delta x_j}{v_{j,q}} \quad (3.3)$$

where

$$v_{j,q} = \frac{1}{\alpha \cdot \frac{1}{V_{j-1,q}} + (1-\alpha) \cdot \frac{1}{V_{j,q}}}$$

3.2.3 Linear Interpolation Between Detectors

Van Lint and van der Zijpp (2003) challenge the constant speed interpolations because they result in instantaneous speed changes where vehicle trajectories would be piecewise linear. A linear speed interpolation between measurement points is proposed (see Fig. 3.3), so that a smooth trajectory is obtained. However, no evidence is presented that drivers behave in this smooth fashion, anticipating slower or faster speed regimes, and driving experience seems to indicate that this is not sound.

The analytic equation for this linear speed interpolation is:

$$v_{j,q}(x) = V_{j-1,q} + \frac{(x - x_{j-1})}{(x_j - x_{j-1})} \cdot (V_{j,q} - V_{j-1,q}) \quad (3.4)$$

And the solution to the differential equation (Eq. 2.2) which defines the virtual vehicle trajectory within a space-time cell (j, q) and an initial condition $(x_{j,q}^0, t_{j,q}^0)$ is expressed as (Van Lint and van der Zijpp 2003):

$$x_{j,q}(t) = x_{j,q}^0 + \left(\frac{V_{j-1,q}}{\Lambda} + x_{j,q}^0 - x_{j-1}\right) \cdot \left(\exp\left[\Lambda\left(t - t_{j,q}^0\right)\right] - 1\right)$$
$$\Lambda = \frac{V_{j,q} - V_{j-1,q}}{x_j - x_{j-1}} \quad (3.5)$$

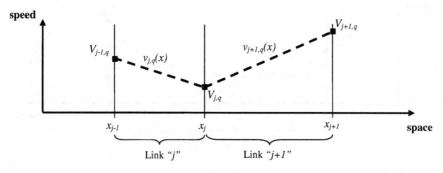

Fig. 3.3 Linear speed trajectory spatial generalization

"Λ" must be significantly greater than zero to avoid numerical problems in the solving of Eq. 3.5. Otherwise, constant speed assumption is equivalent to the linear interpolation, and solutions in Eq. 3.2 can be used.

The trajectory entrance point to the cell depends then on the following condition:

$$x_{j,q}^0 + \left(\frac{V_{j-1,q}}{\Lambda} + x_{j,q}^0 - x_{j-1}\right) \cdot \left(\exp\left[\Lambda\left((q - \Delta t) - t_{j,q}^0\right)\right] - 1\right) < x_{j-1} \quad (3.6)$$

Then:

$$\{x_{j,q}^*, t_{j,q}^*\} = \begin{cases} \left\{x_{j-1}, t_{j,q}^0 + \frac{1}{\Lambda} \ln\left(\frac{\frac{V_{j-1,q}}{\Lambda}}{\frac{V_{j-1,q}}{\Lambda} + x_{j,q}^0 - x_{j-1}}\right)\right\} & \text{if Eq. 3.6 holds} \\ \left\{x_{j,q}^0 + \left(\frac{V_{j-1,q}}{\Lambda} + x_{j,q}^0 - x_{j-1}\right) \cdot \left(\exp\left[\Lambda\left((q - \Delta t) - t_{j,q}^0\right)\right] - 1\right), \; q - \Delta t\right\} & \text{o.w.} \end{cases} \quad (3.7)$$

3.2.4 Quadratic Interpolation Between Detectors

Recently, a quadratic speed interpolation has been proposed by Sun et al. (2008), see Fig. 3.4. This approach tries to mimic the drivers' behavior in relation to speed variations by allowing variable acceleration rates, as drivers may decelerate more when getting close to a congested zone or accelerate more when leaving a congested zone to become free-flow traffic. This method conceptually improves the linear interpolation, in the sense that in the linear case the drivers' behavior excessively anticipates downstream traffic conditions as a result of constant acceleration between measurement points, even before the driver notices the change in the traffic state. Note that in fact, this quadratic approach can be seen as a smoothed approximation to the piecewise constant speed interpolation. However, the problem in this quadratic interpolation is that the "sharp" changes in speed do

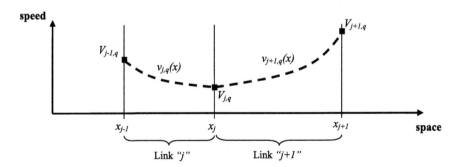

Fig. 3.4 Quadratic speed trajectory spatial generalization

3.2 Methods of Link Travel Time Estimation from Point Speed Measurements

not respond to traffic dynamics or queue evolution but only to the whims of a mathematical function.

The quadratic speed interpolation uses speed observations from three adjacent measurement points. Adapting the formulation presented in Sun et al. (2008) using a Lagrange quadratic interpolation polynomial, the speed trajectory as a function of "x" can be approximated as:

$$v_{j,q}(x) = V_{j-1,q} \cdot \ell_{j-1,q}(x) + V_{j,q} \cdot \ell_{j,q}(x) + V_{j+1,q} \cdot \ell_{j+1,q}(x) \tag{3.8}$$

where the Lagrange basis functions are:

$$\begin{aligned} \ell_{j-1,q}(x) &= \frac{(x - x_j) \cdot (x - x_{j+1})}{(x_{j-1} - x_j) \cdot (x_{j-1} - x_{j+1})} \\ \ell_{j,q}(x) &= \frac{(x - x_{j-1}) \cdot (x - x_{j+1})}{(x_j - x_{j-1}) \cdot (x_j - x_{j+1})} \\ \ell_{j+1,q}(x) &= \frac{(x - x_{j-1}) \cdot (x - x_j)}{(x_{j+1} - x_{j-1}) \cdot (x_{j+1} - x_j)} \end{aligned} \tag{3.9}$$

Equation 3.8 can be rearranged as:

$$v_{j,q}(x) = a \cdot x^2 - b \cdot x + c \tag{3.10}$$

where,

$$\begin{aligned} a &= \frac{V_{j-1,q}}{(x_{j-1} - x_j) \cdot (x_{j-1} - x_{j+1})} + \frac{V_{j,q}}{(x_j - x_{j-1}) \cdot (x_j - x_{j+1})} + \frac{V_{j+1,q}}{(x_{j+1} - x_{j-1}) \cdot (x_{j+1} - x_j)} \\ b &= \frac{V_{j-1,q} \cdot (x_j + x_{j+1})}{(x_{j-1} - x_j) \cdot (x_{j-1} - x_{j+1})} + \frac{V_{j,q} \cdot (x_{j-1} + x_{j+1})}{(x_j - x_{j-1}) \cdot (x_j - x_{j+1})} + \frac{V_{j+1,q} \cdot (x_{j-1} + x_j)}{(x_{j+1} - x_{j-1}) \cdot (x_{j+1} - x_j)} \\ c &= \frac{V_{j-1,q} \cdot x_j \cdot x_{j+1}}{(x_{j-1} - x_j) \cdot (x_{j-1} - x_{j+1})} + \frac{V_{j,q} \cdot x_{j-1} \cdot x_{j+1}}{(x_j - x_{j-1}) \cdot (x_j - x_{j+1})} + \frac{V_{j+1,q} \cdot x_{j-1} \cdot x_j}{(x_{j+1} - x_{j-1}) \cdot (x_{j+1} - x_j)} \end{aligned} \tag{3.11}$$

It can be checked that $b^2 \neq 4ac$. Then, Eq. 3.10 leads to two solutions to differential Eq. 2.2. Firstly, in case that $b^2 > 4ac$:

$$x_{j,q}(t) = \frac{b + \Phi + (\Phi - b) \cdot B \cdot \exp(\Phi \cdot t)}{2 \cdot a \cdot (1 - B \cdot \exp(\Phi \cdot t))} \tag{3.12}$$

where,

$$\Phi = \sqrt{b^2 - 4ac} \tag{3.13}$$

And "B" is the constant obtained applying the initial condition ($x_{j,q}^0$, $t_{j,q}^0$):

$$B = \frac{2 \cdot a \cdot x_{j,q}^0 - b - \Phi}{(\Phi + 2 \cdot a \cdot x_{j,q}^0 - b) \cdot \exp(\Phi \cdot t_{j,q}^0)} \tag{3.14}$$

In this case, the trajectory entrance point to the cell then depends on the following condition:

$$\frac{b + \Phi + (\Phi - b) \cdot B \cdot \exp[\Phi \cdot (q - \Delta t)]}{2 \cdot a \cdot (1 - B \cdot \exp[\Phi \cdot (q - \Delta t)])} < x_{j-1} \tag{3.15}$$

Then:

$$\{x_{j,q}^*, t_{j,q}^*\} = \begin{cases} \left\{ x_{j-1}, \frac{1}{\Phi} \ln\left[\frac{1}{B}\left(\frac{2 \cdot a \cdot x_{j-1} - b - \Phi}{2 \cdot a \cdot x_{j-1} - b + \Phi}\right)\right] \right\} & \text{if Eq. 3.19 holds} \\ \left\{ \frac{b + \Phi + (\Phi - b) \cdot B \cdot \exp[\Phi \cdot (q - \Delta t)]}{2 \cdot a \cdot (1 - B \cdot \exp[\Phi \cdot (q - \Delta t)])}, q - \Delta t \right\} & \text{otherwise} \end{cases} \tag{3.16}$$

Finally, a second solution arises if $b^2 < 4ac$:

$$x_{j,q}(t) = \frac{\Phi' \tan\left(\frac{\Phi' \cdot t}{2} + B'\right) + b}{2 \cdot a} \tag{3.17}$$

where,

$$\Phi' = \sqrt{4ac - b^2} \tag{3.18}$$

And "B'" is the constant obtained applying the initial condition ($x_{j,q}^0$, $t_{j,q}^0$):

$$B' = \tan^{-1}\left(\frac{2 \cdot a \cdot x_{j,q}^0 - b}{\Phi'}\right) - \frac{\Phi' \cdot t_{j,q}^0}{2} \tag{3.19}$$

The trajectory entrance point to the cell in this case depends then on the following condition:

$$\frac{\Phi' \tan\left(\frac{\Phi' \cdot (q - \Delta t)}{2} + B'\right) + b}{2 \cdot a} < x_{j-1} \tag{3.20}$$

Then:

$$\{x_{j,q}^*, t_{j,q}^*\} = \begin{cases} \left\{ x_{j-1}, \frac{2}{\Phi'} \cdot \left[\tan^{-1}\left(\frac{2 \cdot a \cdot x_{j-1} - b}{\Phi'}\right) - B'\right] \right\} & \text{if Eq. 3.20 holds} \\ \left\{ \frac{\Phi' \tan\left(\frac{\Phi' \cdot (q - \Delta t)}{2} + B'\right) + b}{2 \cdot a}, q - \Delta t \right\} & \text{otherwise} \end{cases} \tag{3.21}$$

Note that for particular values of "$V_{j-1,q}$", "$V_{j,q}$" and "$V_{j+1,q}$", the quadratic interpolation "$v_{j,q}(x)$" is not bounded by these measurements. This may result in unrealistic speeds at a particular "x" within the link (i.e. extremely high speeds never measured or extremely low speeds, even negative). This means that the solutions expressed in Eqs. 3.16 and 3.21 cannot be applied directly, and in order to obtain solutions which make sense a truncated definition of the speed evolution must be defined (Sun et al. 2008). Therefore, Eq. 3.8 should be rewritten as:

$$v_{j,q}(x) = \min\left[V_{\max}, \max\left(V_{\min}, V_{j-1,q} \cdot \ell_{j-1,q}(x) + V_{j,q} \cdot \ell_{j,q}(x) + V_{j+1,q} \cdot \ell_{j+1,q}(x)\right)\right] \quad (3.22)$$

where "$V_{\min}$" and "$V_{\max}$" are the speed thresholds to be set.

Given the truncated speed evolution with space within the link, the differential equation (Eq. 2.2) is recommended to be solved numerically.

3.2.5 Criticism to the Presented Methods

The presented speed interpolation models between point measurements have been developed in order to solve the main problem of speed based freeway travel time estimation: the lack of accuracy in case of traffic state transitions within the link.

Constant and piecewise constant models imply instantaneous speed changes which in fact do not occur in real traffic. The remaining approaches seek to obtain continuous speed functions and smoother vehicle trajectories in order to avoid this drawback. For instance, the linear approach distributes the speed change in the traffic transition along the whole link.

However, it is evident from driving experience that traffic state transitions occur in specific spots of the freeway which evolve in time and space at the shockwave speed. When a driver encounters a shockwave, he adapts to the new traffic conditions in a short interval of time and space. This adaptation period depends on the acceleration/braking capabilities of the vehicle, on the driving behavior of the driver (i.e. aggressive or not) and on the perception of accident risk. Either way, it seems evident that the transition will not span for a long time-space period as the linear model assumes, which even implies the driver anticipating the perception of the traffic state change. In order to solve this problem, quadratic interpolations are proposed which imply a more rapid adaptation to speed changes.

None of these advanced methods face the key issue of the problem: where the transition occurs within the link. The proposed mathematical interpolations are blind to traffic dynamics, and hence still prone to errors, as they locate the traffic state transitions according to the whims of the mathematical functions. The improvements in travel times obtained by considering the detailed trajectory of the vehicle within the transition are negligible when compared to the benefits of accurate estimation of the location of the transition at each time period. If there is a situation where these improvements could have a significant contribution, this would be congestion dissolve episodes, where vehicles' acceleration is not so sharp, in relation to the sudden breaking to avoid collision in a congestion onset. In practice, the assumption of instantaneous speed change with the crossing of the shockwave would suffice, as it has been accepted traditionally in the context of continuum traffic flow modeling. Therefore, piecewise constant speed trajectories could be adequate.

These assertions do not imply that the presented constant or piecewise constant models perform better. They are only particular solutions for when the crossing of the shockwave coincides with the speed discontinuity location in the model (e.g. detector location, midpoint …). The piecewise constant speed interpolation method would be adequate provided that the speed change location is accurately estimated (for an online application this is equivalent to a constant weighted speed in the whole section—see Eq. 3.3—where the location of the speed change must be described by an appropriate and dynamic estimation of the parameter "α"). This last issue remains in practice unsolved, as one could employ queuing theory or traffic flow theory to estimate the length of the queue in between detectors, but detector counting errors rapidly accumulate and undermine the results. In specific locations, where a recurrent bottleneck location is detected, one method could be selected among others in relation to the adequacy of its assumptions. On the contrary, in uniform sections, any of the methods will result in the same average errors over sufficiently long time periods

Having said that, the contradictory results found in the literature should not be unexpected as the same method sometimes overestimates travel times and sometimes underestimates them; sometimes considering upstream speed is more accurate and sometimes it is the inverse… It all depends on the location of the traffic transition which evolves with time.

3.3 The Data

In order to provide empirical evidence of the previous statements, it is necessary to compare travel time estimates obtained from average speed data at detector sites with directly measured travel times. Although it is a difficult issue to obtain a representative number of ground truth travel time measurements within each time period "Δt", and for some authors virtually impossible if one wants to consider all the vehicles in a realistic urban freeway (Cortés et al. 2002), the AP-7 turnpike, on the north eastern stretch of the Spanish Mediterranean coast, represents a privileged test site.

The closed tolling system installed on the turnpike, whose objective is to charge every vehicle a particular toll resulting from the application of a kilometric fee to the distance travelled by the vehicle, provides collateral data for every trip on the

3.3 The Data

highway, including the entry junction, the exit junction, and the entrance and exit times. This allows computing origin–destination matrices and travel times between control points on the turnpike (see Chap. 5). In addition, the surveillance equipment installed consists of double loop detectors located approximately every 4 km.

This provides a perfect environment for evaluating travel time estimation methods from loop detector data. The test site, shown in Fig. 3.5, consists of 21.9 km on the southbound direction of the AP-7 turnpike towards Barcelona, Spain. There are 5 detector sites which define 4 links in between. As discussed

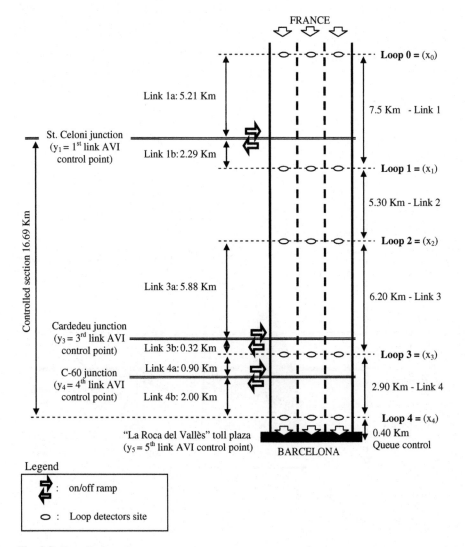

Fig. 3.5 Test site layout

before, the length of these links, ranging from 2.9 to 7.5 km, is far too large in order to consider the practical application of spot speed based travel time estimation methods, as queues take long time to grow over such a long distance, and therefore travel time estimations in transition conditions would be completely flawed. However the behavior of these estimation errors is precisely the issue being analyzed here and their enlargement will be helpful in visualizing the results. Ground truth travel time data are available from control points located by each junction to the downstream exit of the turnpike at "La Roca del Vallès" where the main trunk toll plaza is located. In addition, the queue control system at this main trunk toll plaza provides measured travel times between loop 4 and the 4th AVI control point. This results in a 16.69 km long stretch where loop travel time estimations can be evaluated, as both measured and estimated travel times are available. All data are obtained as 3-min average.

Test data were obtained on Thursday June 21st, 2007, a very conflictive day in terms of traffic in the selected stretch. Problems started around 12:39 when a strict

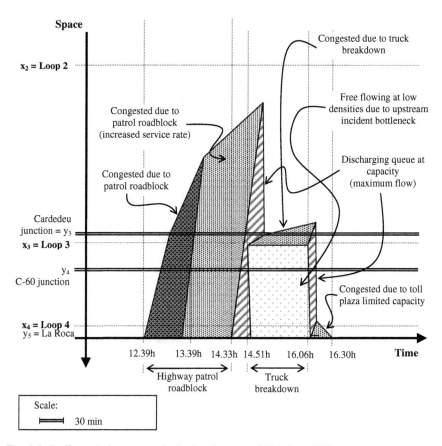

Fig. 3.6 Traffic evolution on test site in the afternoon of 21st June 2007

3.3 The Data

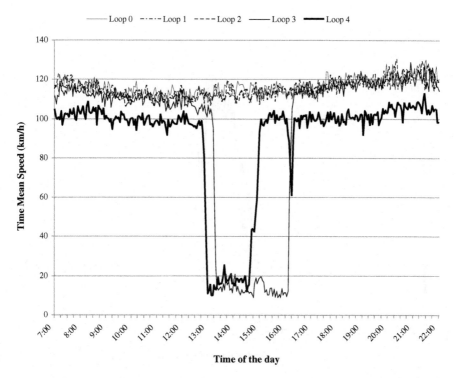

Fig. 3.7 Time mean speed (3 min average) for the whole section measured at loop detector sites

roadblock at "La Roca del Vallès" toll plaza was set up by the highway patrol, reducing its capacity and consequently the output flow. This, in addition to the high traffic demand of the turnpike at that time, caused severe queues to grow rapidly. In view of this fact, and approximately 45 min after the setting of the roadblock, when the queue was already spanning around 5 km, the service rate of the roadblock was increased. This reduced the queue growing rate. Until 14:33 when the patrol roadblock was removed, queues had not started to dissipate. In addition, at 14:51 when queues were still dissipating, the breakdown of a heavy truck within link 4a blocked one out of the three lanes. In turn, this caused queues to start growing upstream again. Finally, when the broken down truck was removed at 16:06, the queues started to dissipate again, flowing at capacity. The high and unanticipated demand at "La Roca" toll plaza, as a consequence of the queue discharge, exceeded the capacity of the toll gates, causing small queues to grow at this location between 16:06 and 16:21.

A complete sketch of traffic evolution on the test site for this particular day can be seen in Fig. 3.6, drawn at scale to match empirical data presented in Figs. 3.7 and 3.8. From Fig. 3.7 it can also be seen how Loop 4 is impacted by the slowing for the toll plaza in free flowing conditions. Also note that the queue never reached Loop 2 location. Besides, Fig. 3.8 shows the low impact of the truck breakdown on vehicles

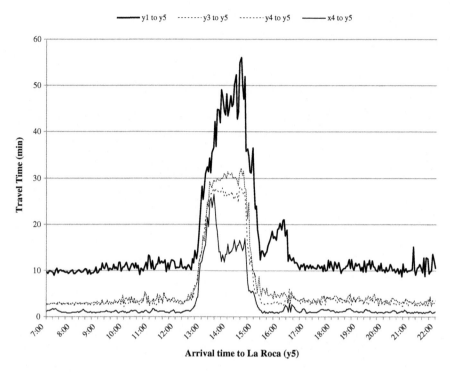

Fig. 3.8 AVI measured arrival based average travel times (3 min average)

entering the highway at "y_3" (i.e. Cardedeu junction), although this entrance was located several hundreds of meters upstream of the induced bottleneck. This is due to the existence of an auxiliary entrance lane at "y_3" which almost allowed bypassing the breakdown location.

3.4 Evaluation of Proposed Speed Spatial Interpolation Methods

Measured AVI travel times shown in Fig. 3.8 are arrival based. Therefore, in order to evaluate the accuracy of different travel time estimations, based on different speed interpolation between point measurements, it is necessary to reconstruct the vehicles' trajectories. The trajectory reconstruction process detailed in Chap. 2 is applied to the speeds field given by the loop test data shown in Fig. 3.7 and considering the space discretization presented in Fig. 3.5 at a time step of 3 min. Once the trajectories have been reconstructed, the resulting travel times are also arrival based, and the comparison with measured travel times is appropriate.

3.4 Evaluation of Proposed Speed Spatial Interpolation Methods

Several methods of speed interpolation between detector sites have been analyzed: constant upstream, constant downstream, midpoint, linear and truncated quadratic (where "V_{min}" = 10 km/h and "V_{max}" = 130 km/h). These methods have been considered as representative of each category, being the results that are easily extrapolated to the remaining methods. By way of illustration, Fig. 3.9 shows the speed profile over space on the test site resulting from the considered interpolation methods.

Obviously, the absolute differences between speed profiles are smaller, as are the differences in the measured speeds. At the limit, they would converge to the same measured speed for the whole corridor. This is reflected in the resulting virtual vehicle reconstructed trajectory, as can be seen in Fig. 3.10. In addition Fig. 3.10 aims to show the behavior of the reconstructed trajectories, for instance piecewise linear in the case of piecewise constant speeds.

The stretch between "Cardedeu junction" ("y_3") and the loop 4 location ("x_4") (see Fig. 3.5 for details) is selected as the evaluation section. This selection responds for several reasons: firstly, ground truth travel times are available for this stretch. Secondly, "y_3" is nearby loop 3 location ("x_3") so that travel time estimations would clearly depend only on speed measurements of loops 3 and 4, making the interpretation of results easier. Finally, afternoon congestion on the test site grows along the whole section so that free flowing traffic, congestion onset, fully congested traffic and congestion dissolve episodes are identifiable. Figure 3.11 plots the comparison between measured and estimated travel times.

From Fig. 3.11, and in accordance with traffic evolution and loop speed measurements presented in Figs. 3.6 and 3.7 respectively, several episodes in relation to the spanning of congestion over the section can be identified:

1. Free flowing traffic for vehicles finishing their trip at "La Roca" between 11:00 and 12:00, among others.
2. Congestion onset between 13:27 and 13:39 (arrival times at "La Roca"), when the queue was growing upstream at an approximated speed of 8.3 km/h. Note from Fig. 3.7 that it took up to 21 min for the queue to grow between "x_4" to "x_3", a 2.9 km section.
3. Congested traffic on the whole stretch between 13:42 and 14:48.
4. Congestion dissolve between 14:51 and 15:00, when the queue was dissolving from downstream due to the removal of the patrol roadblock at an approximated speed of 14.5 km/h. Note from Fig. 3.7 that it took up to 12 min for the queue to dissolve between "x_4" to "x_3".
5. Partially congested stretch due to a lane closure (resulting from the truck breakdown) nearby the upstream end of the stretch between 15:24 and 16:06.

Within episodes "a", "c" and "e", queues do not evolve with time in the evaluation stretch. Small travel time variations are only due to speed variance among drivers in free flowing conditions or to stop&go oscillations in congested traffic. In this case, the average error in the period is a good performance indicator of each travel time estimation method. The average error can be decomposed as a bias (the

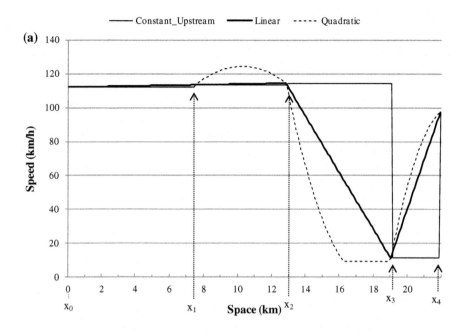

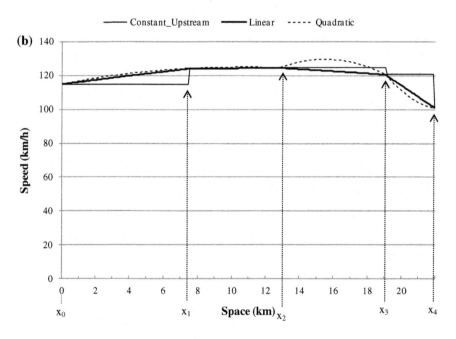

Fig. 3.9 Interpolated speed profiles on the test site. **a** Arrival time 16:00 h—partially congested stretch. **b** Arrival time 20:18 h—free flowing stretch

3.4 Evaluation of Proposed Speed Spatial Interpolation Methods

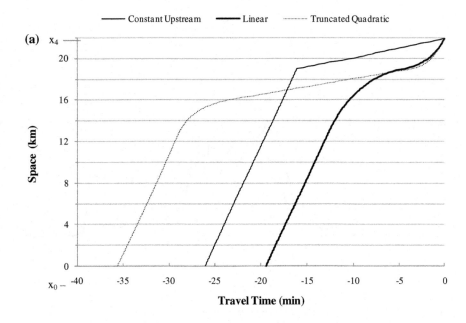

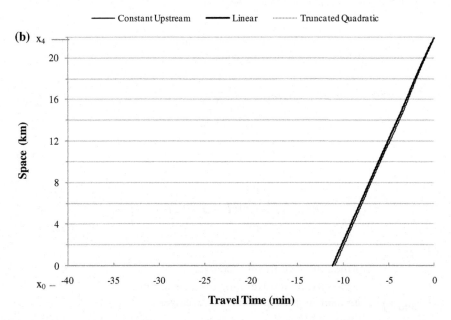

Fig. 3.10 Reconstructed trajectories between "x_4" and "x_0" from different speed interpolations. **a** Arrival time 16:00 h—partially congested trip. **b** Arrival time 20:18 h—free flow trip

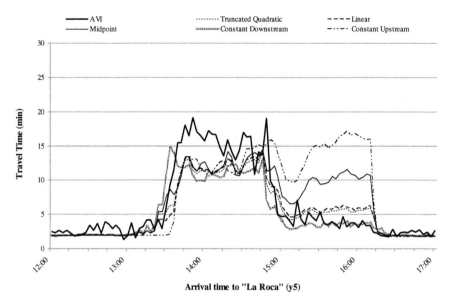

Fig. 3.11 3 min average travel time estimations from reconstructed trajectories on different speed space interpolation assumptions (section between "y_3" and "x_4"; 3.22 km)

mathematical expectation of the error) and a residual (the standard deviation of the error), as formulated in the following equations:

$$RMSE = \sqrt{\frac{1}{n}\sum_{i=1}^{n}(\hat{T}_i - T_i)^2} \quad (3.23)$$

$$Bias = \overline{\hat{T}} - \overline{T} = \frac{1}{n}\sum_{i=1}^{n}\hat{T}_i - \frac{1}{n}\sum_{i=1}^{n}T_i \quad (3.24)$$

$$RRE = \sqrt{\frac{1}{n}\sum_{i=1}^{n}\left[(\hat{T}_i - \overline{\hat{T}}) - (T_i - \overline{T})\right]^2} \quad (3.25)$$

where:
"$\hat{T}_i$"   is the estimated travel time over the section
"T_i"   is the measured travel time over the section
"$\overline{\hat{T}}$" and "$\overline{T}$"   are their respective arithmetic averages

"*RMSE*" stands for the Root Mean Squared Error and "*RRE*" for the Root Residual Error, where:

3.4 Evaluation of Proposed Speed Spatial Interpolation Methods

$$RMSE^2 = bias^2 + RRE^2 \qquad (3.26)$$

Table 3.1 presents these performance indicators for each travel time estimation method in stationary traffic states: free flowing conditions, totally congested stretch and partially congested stretch (i.e. episodes "a", "c" and "e").

Qualitatively in Fig. 3.11 and quantitatively in Table 3.1, it is shown that the performance of all methods is almost the same in episodes where uniform traffic conditions span for the whole section (episodes "a" and "c").

All methods perform well in free flowing conditions, as the average error is of the same order of magnitude as the travel time standard deviation. However, a systematic underestimation is observed. This results from a speed overestimation at loop detector sites in free flowing conditions in relation to the average speed of vehicles across the turnpike section. Two reasons explain this bias: One, the computations of time mean speeds instead of the lower space mean speeds, as stated in the introduction of the chapter; Two, loop detectors are installed on privileged spots of the highway, far from problematic sections where speed drops off, like junctions and weaving sections.

On the contrary, all methods are not accurate enough when congestion spans for the whole stretch, reporting average errors at approximately twice the standard deviation of travel times in these episodes. Again, systematic underestimation is observed. Computing space mean speeds would slightly improve this bias, but it

Table 3.1 Numerical differences between measured and estimated travel times for an aggregation period of 3 min on the section between "y_3" and "x_4" (3.22 km)—stationary conditions

(a)			AVI measured travel time		Constant upstream estimation		Constant downstream estimation	
			Mean	Dev.	Bias	RRE	Bias	RRE
a) Free flowing		min	2.11	0.62	−0.38	0.59	−0.20	0.59
		%	−	−	−18	28	−10	28
c) Totally congested		min	15.67	1.97	−3.08	2.50	−4.66	2.73
		%	−	−	−20	16	−30	17
e) Partially congested		min	3.75	0.66	11.95	1.22	−0.27	0.77
		%	−	−	318	33	−7	20
(b)			Midpoint estimation		Linear estimation		Truncated quadratic estimation	
			Bias	RRE	Bias	RRE	Bias	RRE
a) Free flowing		minutes	−0.29	0.59	−0.29	0.59	−0.27	0.59
		%	−14	28	−14	28	−13	28
c) Totally congested		minutes	−3.34	2.51	−3.69	2.45	−3.87	2.52
		%	−21	16	−24	16	−25	16
e) Partially congested		minutes	6.66	1.00	1.97	0.74	1.68	0.74
		%	178	27	52	20	45	20

must be noted that the main reason for this underestimation is the biased speed measurement of loop detectors in stop&go situations, when only the "go" part of the movement is measured.

A different behavior of the methods appears in the case of a partially congested stretch (where the part covered by the queue does not evolve with time). Note that episode "e" corresponds to a situation where the queue only spans for a few upstream meters of the section, but stepping on the loop spot. These results in travel times are similar to the free flowing situation, but very low speeds are measured at the upstream loop location. It should be clear in this situation why constant downstream estimation outperforms all the other methods, and why constant upstream produces completely flawed travel times. Obviously, midpoint, linear and truncated quadratic estimations are in between. It is interesting to note that linear and truncated quadratic estimations beat midpoint estimations. Recall from Sect. 2.2 that piecewise constant approaches are equivalent to travel times resulting from a constant weighted harmonic average speed where the weighting factors are the relative coverage of each piece. Linear and truncated quadratic estimations could also be seen as piecewise constant approaches, with infinitesimally small pieces. One can easily realize that, while the arithmetic average of speeds would be approximately equal in all three approaches, harmonic ones are not, due to a higher influence of lower speeds. This is the reason why midpoint travel time estimations are significantly higher than linear and quadratic ones.

This does not mean that linear and truncated quadratic approaches outperform the midpoint algorithm in partially congested situations. It would be true in a case where the queue covers a small part of the section (like the situation analyzed here), but it would be the inverse if the queue spans for almost the whole length of the stretch.

What is evident from the presented results is that none of the methods are intrinsically better than the other in the case of partially congested situations. The best one is dependent on matching the method assumptions with the queue coverage of the section.

This evidence is also seen in situations when queues evolve with time within the section. Episode "b" corresponds to a congestion onset from downstream (congestion grows against traffic direction) while during episode "d" congestion dissolves also from downstream (against traffic direction). Table 3.2 presents numerical results of the absolute errors committed with each estimation approach, and shows how the error decreases when traffic state approaches the assumptions of the method. The logic of the absolute error behavior is clear, although some values may seem on the wrong side, resulting from the systematic overestimation of average speeds in congested conditions (see Table 3.1). Figure 3.12 may help in the interpretation of numerical values in Table 3.2.

Table 3.2 Numerical differences between measured and estimated travel times for an aggregation period of 3 min on the section between "y_3" and "x_4" (3.22 km)—evolving conditions

	Arrival time at "La Roca"	AVI measure (min)	Absolute error (min)				
			Constant upstream	Constant downstream	Midpoint	Linear	Truncated quadratic
b) Congestion onset	13:27	2.87	−1.07	3.49	2.50	1.27	2.06
	13:30	5.73	−3.89	5.05	1.91	−1.76	−0.54
	13:33	9.12	−7.11	5.68	−0.50	−4.46	−2.87
	13:36	11.34	−7.69	2.60	−3.46	−6.08	−5.24
	13:39	15.39	−6.36	−3.42	−6.42	−6.20	−6.08
d) Congestion dissolve	14:48	18.96	−3.94	−12.13	−7.72	−8.76	−9.57
	14:51	9.77	5.99	−4.06	1.71	−0.86	−1.71
	14:54	8.72	6.61	−2.65	3.35	0.26	−0.34
	14:57	4.89	8.80	−0.04	5.41	3.37	2.75

3.5 Conclusions and Further Research

There is a need for travel time information in metropolitan freeways, where in most cases solely loop detector surveillance is available. Several methods have been developed in order to estimate travel times from speed measurements at loop detector sites whose main differences lie in the speed interpolation approach between point measurements. In fact, the ignorance of speed evolution between measurement points represents a major drawback for these types of methods.

The present chapter demonstrates conceptually and with an accurate empirical comparison resulting from accurate travel time definitions, that travel time estimation methods based on mathematical speed interpolations between measurement points, which do not consider traffic dynamics and the nature of queue evolution, do not contribute in an intrinsically better estimation, independently of the complexity of the interpolation method. All of them show a similar performance when a unique traffic state covers the whole target stretch. It can be concluded that all methods perform well in free flowing conditions in spite of a slight systematic underestimation. In contrast, all methods provide highly underestimated estimations in completely congested sections, resulting in unrealistic travel times. The main reason for this bad performance is the inability of loop detectors to capture the speed oscillations produced by stop&go traffic, resulting in inaccurate and systematic overestimated average speed estimates. The improvement of average speed loop measurements under congested situations should be considered as an issue for further research.

Major errors can arise in the case of partially congested sections, resulting from a congestion onset (or dissolve) episode or due to the activation of a bottleneck within the section (either recurrent or incident related). In this situation, different methods provide significantly different estimations. The quality of each method relies on the fitness of the method assumptions of the real evolution of the queue along the

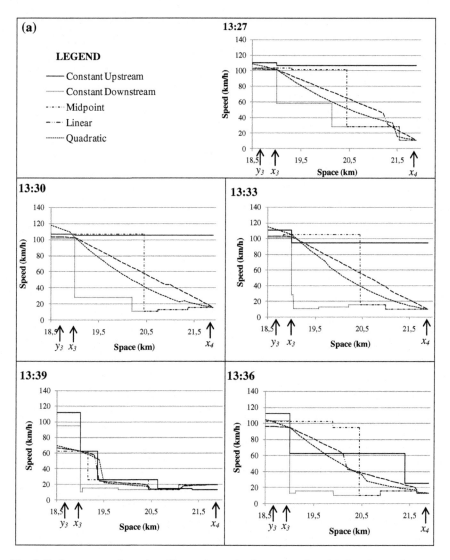

Fig. 3.12 Reconstructed speed profiles on the section between "y_3" and "x_4". **a** Congestion onset episode. **b** Congestion dissolve episode

section. In practice, the relative benefits between commonly used interpolation assumptions, like constant speed over the whole stretch, arbitrary location for the speed change (e.g. midpoint algorithm), or speed profiles resulting from mathematical interpolations blind to traffic dynamics (e.g. linear or truncated quadratic approaches) are site specific, and in the case of indiscriminate use, depend on chance. Avoiding methods based on only one loop detector measurement (e.g. constant upstream or constant downstream) will prevent the highest punctual errors.

3.5 Conclusions and Further Research

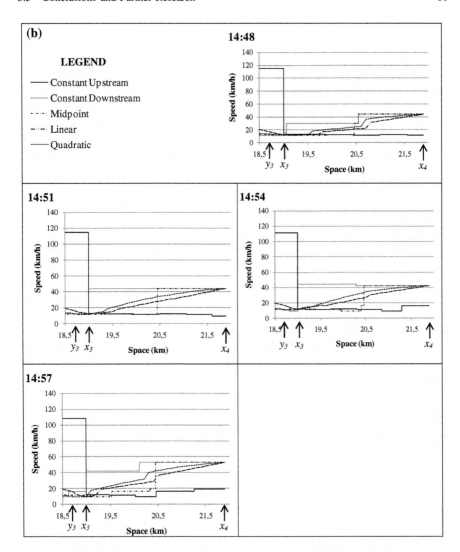

Fig. 3.12 (continued)

Therefore, under no better approach, midpoint, linear or truncated quadratic methods are recommended, from which midpoint algorithm stands out for its simplicity. The key issue for an efficient method should be the estimation of queue length within the section so that each speed measurement could be assigned to an adequate length of the highway stretch.

The absolute magnitude of these estimation errors directly depend on the loop detector spacing. Lower surveillance densities result in higher probability of a traffic transition within the section defined by two consecutive loops. In addition, the estimation error would be propagated over a longer length, resulting in higher absolute errors. Being aware of this situation, and realizing the difficulties of locating traffic state transitions in long sections, most traffic agencies have chosen to invest in higher loop densities (typically 1 loop detector every 500 m), in order to obtain realistic travel times from their measurements. In this configuration, based on intensive surveillance, the selected travel time estimation method does not matter. However, the frequent detector failure and malfunctioning, which translates to temporally increased detector spacing, should also be taken into account.

The conclusions of this chapter should not be taken as suggesting not to use the existing schemes in surveillance configurations where this intensive monitoring is not available, but that they should be used only carefully and not be taken as perfect. For instance, most of the methods can be used to provide an upper and lower bound on the travel time, using one end or the other of a link.

In a real-time context, the main advantage of travel times estimated from loop measurements is the possibility of obtaining true travel times; this is to obtain a virtual measurement for a vehicle travel time before the end of its journey. This provides benefits in the immediacy of the detection of transitions in the traffic stream state.

From the chapter it is concluded that on the one hand, directly measuring travel times provides accuracy benefits, although delayed information. On the other hand, the indirect estimation from speed measurements provides immediacy in exchange for a loss of accuracy. It seems that both types of measurements should be complementary. A data fusion scheme capable of taking the better of each one seems an appealing alternative.

References

Abdulhai, B., & Tabib, S. M. (2003). Spatio-temporal inductance-pattern recognition for vehicle reidentification. *Transportation Research Part C, 11*(3–4), 223–239.

Claudel, C. G., Hofleitner, A., Mignerey. N., & Bayen, A. M. (2009). *Guaranteed bounds on highway travel times using probe and fixed data.* Paper presented in the 88th Transportation Research Board Annual Meeting, Washington D.C.

Coifman, B. (2001). Improved velocity estimation using single loop detectors. *Transportation Research Part A, 35*(10), 863–880.

Coifman, B. (2002). Estimating travel times and vehicle trajectories on freeways using dual loop detectors. *Transportation Research Part A, 36*(4), 351–364.

Coifman, B., & Cassidy, M. (2002). Vehicle reidentification and travel time measurement on congested freeways. *Transportation Research Part A, 36*(10), 899–917.

Coifman, B., Dhoorjaty, S., & Lee, Z. (2003). Estimating the median velocity instead of mean velocity at single loop detectors. *Transportation Research Part C, 11*(3–4), 211–222.

Coifman, B., & Ergueta, E. (2003). Improved vehicle reidentification and travel time measurement on congested freeways. *ASCE Journal of Transportation Engineering, 129*(5), 475–483.

Coifman, B., & Krishnamurthya, S. (2007). Vehicle reidentification and travel time measurement across freeway junctions using the existing detector infrastructure. *Transportation Research Part C, 15*(3), 135–153.

References

Cortés, C. E., Lavanya, R., Oh, J., & Jayakrishnan, R. (2002). General-purpose methodology for estimating link travel time with multiple-point detection of traffic. *Transportation Research Record: Journal of the Transportation Research Board, 1802*, 181–189.

Dailey, D. J. (1993). Travel time estimation using cross-correlation techniques. *Transportation Research Part B, 27*(2), 97–107.

Dailey, D. J. (1999). A statistical algorithm for estimating speed from single loop volume and occupancy measurements. *Transportation Research Part B, 33*(5), 313–322.

Fariello, B. (2002). *ITS America RFI travel time projects in North America*. San Antonio TransGuide Program, Texas Department of Transportation.

Hellinga, B. (2002). Improving freeway speed estimates from single loop detectors. *ASCE Journal of Transportation Engineering, 128*(1), 58–67.

Kothuri, S. M., Tufte, K. A., Ahn, S., & Bertini, R. L. (2007). *Using archived ITS data to generate improved freeway travel time estimates*. Paper presented in the 86th Transportation Research Board Annual Meeting, Washington D.C.

Kothuri, S. M., Tufte, K. A., Fayed, E., & Bertini, R. L. (2008). Toward understanding and reducing errors in real-time estimation of travel times. *Transportation Research Record: Journal of the Transportation Research Board, 2049*, 21–28.

Kwon, E. (2004). Development of operational strategies for travel time estimation and emergency evacuation on freeway network. Final Report MN/RC – 2004-49. Prepared for the Minnesota Department of Transportation.

Li, R., Rose, G., & Sarvi, M. (2006). Evaluation of speed-based travel time estimation models. *ASCE Journal of Transportation Engineering, 132*(7), 540–547.

Lin, W., Dahlgren, J., & Huo, H. (2004). Enhancement of vehicle speed estimation with single loop detectors. *Transportation Research Record: Journal of the Transportation Research Board, 1870*, 147–152.

Lucas, D. E., Mirchandani, P. B., & Verma, N. (2004). Online travel time estimation without vehicle identification. *Transportation Research Record: Journal of the Transportation Research Board, 1867*, 193–201.

Mikhalkin, B., Payne, H., & Isaksen, L. (1972). Estimation of speed from presence detectors. *Highway Research Board, 388*, 73–83.

Nam, D. H., & Drew, D. R. (1996). Traffic dynamics: Method for estimating freeway travel times in real time from flow measurements. *ASCE Journal of Transportation Engineering, 122*(3), 185–191.

OECD/JTRC. (2010). *Improving reliability on surface transport networks*. Paris: OECD Publishing.

Oh, J. S., Jayakrishnan, R., &Recker, W. (2003). *Section travel time estimation from point detection data*. Paper presented in the 82nd Transportation Research Board Annual Meeting, Washington D.C.

Palen, J. (1997). The need for surveillance in intelligent transportation systems. *Intellimotion, 6*(1), 1–3, 10 (University of California PATH, Berkeley, CA).

Petty, K. F., Bickel, P., Ostland, M., Rice, J., Schoenberg, F., Jiang, J., & Rotov, Y. (1998). Accurate estimation of travel time from single loop detectors. *Transportation Research A, 32*(1), 1–17.

Pushkar, A., Hall, F., & Acha-Daza, J. (1994). Estimation of speeds from single-loop freeway flow and occupancy data using cusp catastrophe theory model. *Transportation Research Record: Journal of the Transportation Research Board, 1457*, 149–157.

Rakha, H., & Zhang, W. (2005). Estimating traffic stream space-mean speed and reliability from dual and single loop detectors. *Transportation Research Record: Journal of the Transportation Research Board, 1925*, 38–47.

Soriguera, F., & Robusté, F. (2011). Estimation of traffic stream space-mean speed from time aggregations of loop detector data. *Transportation Research Part C, 19*(1), 115–129.

Sun, L., Yang, J., & Mahmassani, H. (2008). Travel time estimation based on piecewise truncated quadratic speed trajectory. *Transportation Research Part A, 42*(1), 173–186.

Treiber, M., & Helbing, D. (2002). Reconstructing the spatio-temporal traffic dynamics from stationary detector data. *Cooperative Transportation Dynamics 1*, 3.1–3.24.

Turner, S. M., Eisele, W. L., Benz, R. J., & Holdener, D. J. (1998). *Travel time data collection handbook*. Research Report FHWA-PL-98-035. Texas Transportation Institute, Texas A&M University System, College Station, Tex.

van Lint, J., & van der Zijpp, N. (2003). Improving a travel-time estimation algorithm by using dual loop detectors. *Transportation Research Record: Journal of the Transportation Research Board, 1855*, 41–48.

Wang, Y., & Nihan, L. H. (2000). Freeway traffic speed estimation with single-loop outputs. *Transportation Research Record: Journal of the Transportation Research Board, 1727*, 120–126.

Wang, Y., & Nihan, L. N. (2003). Can single-loop detectors do the work of dual-loop detectors? *ASCE Journal of Transportation Engineering, 129*(2), 169–176.

Chapter 4
Design of Spot Speed Methods for Real-Time Provision of Traffic Information

Abstract The accuracy of travel time information disseminated in real-time on metropolitan freeways is one of the key issues in the development of advanced traveler information systems. It is generally considered that travel time estimations based on spot speed measurements at loop detectors are not accurate enough to support this real-time information need. This has brought traffic agencies to the view that for real-time information systems to be effective they would need to have much more accurate ways of measuring travel times. Many fancy technologies to directly measure vehicular travel times are being proposed. This chapter shows that, in the real-time context, the precision of the system is not related solely to the accuracy of the measurement. Immediacy and forecasting capabilities play a role. Therefore, focusing only on the accuracy of the travel time measurement is a myopic approach, which can lead to counterintuitive results. Specifically, it is claimed that using travel times estimated with the traditional spot speed Midpoint Algorithm, the performance of the real-time information system is better than by using much more accurate directly measured travel times. Guidelines for an adequate configuration of the common parameters of the system are provided. In addition, real-time context enhancements for travel time estimation methods based on punctual speed measurements are proposed. These are addressed by taking into account an easy and practical implementation. They have been proven to work well in an empirical application on a Spanish Freeway.

Keywords Freeway travel time · Real-time information systems · Loop detector data

4.1 Introduction

Information on the expected travel time along a congested freeway corridor ("expected" must be stressed as it will play a significant role) is perhaps the most valuable traffic information for drivers in order for them to improve the quality and efficiency of their trips (Palen 1997). Pre-trip information may allow drivers the selection of departure time and route or even a mode shift. On-trip information is valuable for rerouting or deciding to accept park&ride options. In both situations, travel time information contributes to congestion mitigation. Even in the case where no travel time improvement were possible, travel time information would still improve the quality of the journey by reducing the uncertainty and consequently the stress of the driver. It must be highlighted that the accuracy of the disseminated information is crucial, as providing inaccurate travel time estimations can be detrimental.

Note that travel time information is requested before the trip starts (pre-trip) or while traveling (on-trip). In contrast, the trip must be finished in order to measure its travel time. The requested information is therefore a prediction. Clearly two contexts may be identified for this predicted information. Firstly, when travel time information is requested a "long" time before the trip is intended (i.e. at least several hours in advance and clearly pre-trip). Secondly, when travel time information is to be provided just when starting the trip or while traveling (i.e. on-trip or very short pre-trip information). This last context will be referred to as real-time information. These two contexts are radically different. The decisions to be made by the user are different, the information requirements are different (and so will be the estimation methods) and even the technologies to disseminate the information are different (OECD/JTRC 2010). For instance, in the case of "long term" travel time prediction, users seek (for example on the web) information to support a trip planning decision (e.g. mode, route, departing time) at a strategic level. The user wants to gain the knowledge that a frequent traveler would have on the intended trip (note that sporadic travelers are more prone to request long term travel time predictions, as they obtain more value from them). In this case the useful information to be disseminated includes the average recurrent behavior of travel times on the selected route and some indication of its reliability (e.g. frequency of non-recurrent incidents). This information could be obtained from an adequate treatment of historical travel time measurements from which patterns could be derived (Soriguera and Rosas 2012), the current traffic conditions being almost meaningless. The user accepts the information belonging to a most likely scenario, and is aware of the existing probability of something happening and the prediction being wrong.

In contrast, real-time information should allow the driver to make an immediate operational decision regarding his trip (e.g. to use or not to use park&ride/transit options, to change or not to change to an alternative route, to reschedule or not reschedule the tasks at his destination, or even to decide to give up the travel intention). Although the requested real-time information is still a prediction (i.e. a short term prediction where the horizon is precisely the expected travel time), the driver's expectation is that the information provided (for instance by means of Variable Message Signs—VMS—smartphones or network connected on-board navigation devices) is almost certain. This real-time travel time information, which includes the possible non-recurrences and which is equally valuable for sporadic and frequent travelers, must be based mostly on the measurement of current traffic conditions. The historical long term patterns would add very little value in this case.

4.1 Introduction

In a perfectly idealized and deterministic world (which by no means implies neither a perfect nor an ideal world—the beauty of randomness—but surely an easier to manage one), the current traffic state could be exactly measured (i.e. initial conditions). Short-term future inputs to the system (i.e. traffic demand) would also be known in detail. And finally, the relationship between infrastructure performance and driver behavior (i.e. the traffic model) would be invariable, exact and known. In this idealistic world, given that the initial conditions, the current and future demands and the traffic model are exactly and deterministically known, real-time information would be predicted with certainty.

Despite traffic engineer attempts to assimilate freeways to the deterministic environment described (e.g. enhancing traffic surveillance equipment, controlling demand at on-ramps, imposing more strict traffic regulations), we are still far off from this ideal situation. And we will continue to be to a certain degree while travel decisions are freely made and drivers are human beings. This means that current traffic state conditions are not perfectly known. Measurement is usually limited to some discrete control points. Future demand can be grossly predicted, but not exactly known in advance. Driver behavior and freeway performance models are approximate, and non-recurrent incidents are unpredictable.

Surely, and taking into account the intensive research on all these topics, we will come to a situation where the combination of monitoring, knowledge and control measures will result in a certain real-time freeway travel time information, in relation to the decisions to be supported. Meanwhile, however, traffic engineers are looking for a practical solution that should be based on actual surveillance equipment installed (i.e. economically feasible), that should provide an acceptable estimation (i.e. still valuable information despite its inaccuracy) and that is preferably simple. Given the requirements of this provisional solution and the degree of uncertainty in all parts of the problem (i.e. initial conditions, demand evolution and traffic model), it seems an adequate approach to deal jointly with all these uncertainties in a way that one counterbalances the others. This is precisely the objective of the present chapter. In contrast, a solution based only on a more precise estimation of one of the parts of the problem, which could be a building block for a future more smart solution, would be clearly myopic in an actual implementation.

In this chapter it is shown that this practical simple solution exists, and could be considered as a business as usual approach. It is proven that the traditional travel estimation methods based on loop detector punctual speed measurements (see Chaps. 1 and 3 partially devoted to the description of travel time estimation methods), which are considered highly inaccurate (see Chap. 3) and whose use have been discouraged, can in fact provide more reliable real-time information than other more accurate measurement techniques. This is an example where more accuracy in the measurement can reduce precision in the final objective.

A conceptual proof of this claim is presented in Sect. 4.2, which makes use of the basic insights in travel time definitions discussed previously in Chap. 2. Section 4.3 proposes some simple improvements to the traditional methods to better feed real-time needs. Finally, Sects. 4.4 and 4.5 are devoted to the empirical verification of the previous claims using data obtained on a metropolitan freeway near Barcelona, Spain. Some final conclusions are presented in Sect. 4.6.

4.2 Design of Spot Speed Methods for Real-Time Provision of Traffic Information

The common practice in the real-time implementation of travel time information systems based on point speed measurements (which are the great majority) is to estimate link "$T_T(A)$" by means of the available loop detectors (using one of the methods described in Chap. 3) and add them up to obtain the corridor travel time, "$T_T(C)$". This is the information disseminated in real time. Being aware that spot speed travel time estimation methods are highly inaccurate in some situations (see Table 4.1) traffic managers blame themselves for not being able to more accurately measure freeway travel times. They would be glad to have (and are wishing to invest in) means of directly measuring travel times or ways to track the queue evolution within freeway sections in order to improve the accuracy of the spot speed estimation. If they succeed in this objective, they would realize with disbelief that in some situations the performance of the real time information system does not improve with the accuracy of the measurement. On the contrary, its performance is worsened.

In the context of real time information systems, "goodness" of the estimation should be defined as how the disseminated information (i.e. the measured "$T_A(C)$", or the estimated "$T_T(C)$") approach the desired "$T_D(C')$" (where C' stands for time—space zone defined by the whole corridor and time interval "$p + 1$"). Clearly two situations can be defined. Firstly the situation where "$T_A(C) = T_T(C) = T_D(C')$". This happens when travel times do not evolve with time or evolve very slowly (i.e. approximate time stationary conditions). In this situation the performance of the real-time information system is directly related to the accuracy of the measurement. The real-time requirement does not play any differential role in this case. In practice, little can be done to improve the accuracy of spot speed travel time estimation methods in this case (see Table 4.1). Loop detectors are already installed and their locations must be assumed fixed. Computation procedures at roadside controllers also are difficult to modify in practice. Even more difficult would be for the loop measurements to account for stop&go oscillations. Only some gains can be achieved by time smoothing the speed measurements. Care must be taken to only apply the smoothing process when approximate time stationary conditions prevail. An intelligent smoothing process that accounts for this fact is presented in the next section.

The second situation arises when traffic conditions are evolving rapidly with time and within the corridor. In this case, considering typical corridors of several km in length, "$T_A(C) \neq T_T(C) \neq T_D(C')$". Then the quality of the real time travel

4.2 Design of Spot Speed Methods for Real-Time Provision …

Table 4.1 Performance of spot speed travel time estimation methods in different traffic conditions

Traffic state within the section (defined by 2 consecutive detectors)	Spot speed methods	Spot speed methods performance	Suggested directions for improving measurement accuracy		Real-time information system: practical approach
Approx. stationary time-space conditions	Whole free flowing section	Good Slight systematic underestimation (14 %)	• Compute space-mean speeds • Detectors to be installed in representative spots		Use Midpoint Algorithm • Improve accuracy (do whatever you can from previous column) • Intelligent smoothing process • Data fusion scheme when possible
	Whole congested section	Medium-bad Significant underestimation (24 %)	• Improve loop speed measurement in stop&go conditions. • Average speed measurements over longer time intervals		
Approx. time stationary conditions	Partially congested section (stationary)	Very bad Context dependent	• Locate the inner section bottleneck and divide the section accordingly	• Increase loop detector density (it reduces the average magnitude of the error and the relative number of affected sections)	
Rapidly evolving conditions	Partially congested section (non stationary)	Very bad	• Track the back of the queue and use the dynamically divided section accordingly		Use Midpoint Algorithm Improving accuracy can be detrimental

Note In accordance with results presented in Chap.3

time information does not only depend on the accuracy of the measurement but also on the immediacy in reporting the information and on its forecasting capabilities.

The immediacy requirement accounts for the error resulting from "$T_A(C)$" or "$\hat{T}_T(C)$" (i.e. the spot speed estimation of "$T_T(C)$"), being different from the actual true travel times, "$T_T(C)$". In case of using "$T_A(C)$", the immediacy error results from the delay in the information. For "$\hat{T}_T(C)$" it is, in fact, a measurement error due to limited measurement points, that could be eliminated in the idealistic case of perfect surveillance.

In addition, the required forecasting capabilities account for the error incurred due to the future evolution of traffic conditions in the real-time information forecasting horizon (i.e. the difference between "$T_T(C)$" and "$T_D(C')$"). This error has a higher degree of uncertainty.

It is shown next that, due to its immediacy, a spot speed estimated "$\hat{T}_T(C)$" is preferable to an accurately measured but outdated "$T_A(C)$". It is also shown that the lack of accuracy of spot speed methods when rapidly evolving traffic conditions prevail is compensated by their forecasting capabilities. In these situations, increasing the accuracy of the measurement may imply a reduction of the forecasting capabilities of the method, so that the overall result could be counterproductive.

4.2.1 Immediacy Requirement in Rapidly Evolving Traffic Conditions

On a real time basis, and given two equally accurate travel time estimations, the one which provides more immediacy is preferable. Assuming that "$\hat{T}_T(C)$" and "$T_A(C)$" are available, it is necessary to compute the expected delay in the information in each case to obtain a supported selection. The effects of "Δt" are neglected, as they would affect equally both types of estimation.

The rapid traffic state transitions involved in the situations under study can only respond to 3 typologies, (a), (b) and (c):

(a) *Congestion onset.* Congestion always grows from downstream and against the traffic flow direction. The rare moving bottleneck episodes that would contradict this statement can be neglected.
(b) *Congestion dissolves from upstream.* And in the same direction as traffic. This type of transition arises in case of a reduction in freeway demand.
(c) *Congestion dissolves from downstream.* And against traffic direction. This type of transition arises in case of an increase in the bottleneck capacity (e.g. the incident is removed).

The information delay in reporting the change in traffic conditions for "$\hat{T}_T(C)$" based on loop detector measurements is equal to the time elapsed between the transition start and reaching the detector location. This clearly depends on the detector density. In the worst case, this delay would be equal to the time required by the traffic transition to cover the whole section length, "$\Delta x/u$", being "u" the transition speed (known as the "shock" speed in the context of traffic dynamics). It can be derived from continuum traffic flow theory that "$u = \Delta q/\Delta k$", the ratio of the difference in flow over the difference in density between the intervening traffic states. From the realistic triangular shape assumption for the freeway fundamental diagram [i.e. the relationship between flow (*q*) and density (*k*); for a detailed explanation on all these topics see Daganzo (1990)], it follows that all the possible shock speeds involved in transitions of type (*a*) are bounded between [−*w*, 0] being extreme speeds very rare. "−*w*" is the shock speed between any two congested traffic states, or equivalently the slope of the congested branch of the triangular fundamental diagram. In freeways, it can be assumed to be somewhere between

−15 and −20 km/h. Similarly, shock speeds involved in transitions of type (*b*) are bounded between [0, v_f] again being extreme speeds theoretically possible but very rare in the real world. "v_f" stands for the free flowing speed on the freeway. Finally, as active bottlenecks discharge at capacity, shock speeds in transitions type (*c*) will evolve at "−*w*" speed.

However, if only focusing on drastic and sharp traffic transitions, very low speeds (in absolute magnitude) do not need to be considered. A 5 km/h lower bound in the absolute magnitude of the shock speed could be selected to define these rapidly evolving transitions. In this case, and taking into account that "Δx" must be short to avoid large measurement errors, the worst case information delay for these methods will in most situations be much lower than the information delay in the case of using "$T_A(C)$", which is equal to the travel time experienced by the drivers while traversing the congested portion of the freeway and reaching the final target destination. This can be seen by realizing that usually, the corridor length is much longer than the section length (i.e. $\sum_{j \in C} \Delta x_j \gg \Delta x_j, \forall j \in C$) and that the order of magnitude of congested freeway speeds is the same as the shock speeds.

However, it follows from this discussion that for short corridor lengths (e.g. let's say <15 km) or for longer ones when the incident happens on its downstream part, in the very early development of congestion, the information delay incurred by using "$T_A(C)$" is similar to that using "$\hat{T}_T(C)$" (because under these conditions the time elapsed between vehicles entering the congestion and reaching the corridor end—the "$T_A(C)$" information delay—is short). This means that both types of information provide similar immediacy in reporting the start of a congestion period. However, when congestion grows, the previous statement is no longer true. Then the evolution of this period is more rapidly reported using "$\hat{T}_T(C)$".

One assumption has been considered: "Δx" is short in order to avoid large measurement errors. The average measurement error (assuming the transition eventually covers the whole section and uniform probability of incidents in each spot within the section) will be equal to:

$$|\varepsilon_{measurement}| \approx \frac{\Delta x}{2} \cdot \left[\frac{1}{v_{congestion}} - \frac{1}{v_f}\right] \quad (4.1)$$

$$\Delta x \leq \frac{2 \cdot v_{congestion} \cdot v_f \cdot |\varepsilon_{max-measurement}|}{v_f - v_{congestion}} \quad (4.2)$$

Plugging in typical congested and free flowing average freeway speeds, and considering a maximum error of 1 min in the transitional section (adequate considering that the granularity of the disseminated information will not be lower than that), Eq. 4.2 results in maximum "Δx" of the order of 0.5–2 km. This high detector density is only required in frequently congested freeway sections.

4.2.2 Midpoint Algorithm Forecasting Capabilities

It has been shown that "$\hat{T}_T(C)$" provides the desired immediacy in reporting any travel time variation. But this only solves half the problem, as traffic will evolve from the instant of the measurement to the time the traveler actually undertakes the journey in the next time interval. This section is devoted to show how the "forecasting" capabilities of the rough "$\hat{T}_T(C)$" estimation obtained with the Midpoint Algorithm, makes it preferable to a more accurate "$T_T(C)$" measurement. Possible applications of other constant speed methods will be discussed next.

The Midpoint Algorithm deals with uncertainties in the measurement in a way that moves "$\hat{T}_T(C)$" closer to the desired "$T_D(C')$". This concept is sketched in Table 4.2, where the most probable current traffic state at the instant the transition is detected can be seen. The section containing the bottleneck is neglected. This is not a significant drawback because it will not be the most representative section, as it is only one amongst several of them. In addition, this specific case is analyzed further in the next section devoted to stationary transitions. Table 4.2 also shows the most probable traffic state that will prevail in the short term. The assumption here is that the nature of the transition will not change in the immediate future. Obviously this assumption will fail sometimes, but it is the best one can make while still avoiding the uncertainties of forecasting.

Note that an accurate estimation of "$T_T(C)$" would report the travel times corresponding to the current traffic state, while the Midpoint Algorithm estimation, "$\hat{T}_T(C)$", reports a travel time estimation which approaches the direction of most probable future traffic states. This is naïvely defined as the forecasting capabilities of the Midpoint Algorithm.

The dependency of these forecasting capabilities on "Δx" is clear. Longer sections imply a longer horizon of the prediction (because it will take more time for the

Table 4.2 Traffic transitions under constant evolution assumption

Type of traffic transition	Most probable current traffic state when transition is detected	Most probable future traffic state
(a) Congestion onset		
(b) Congestion dissolve from upstream		
(c) Congestion dissolve from downstream		
Legend: ▨ Congested freeway portion ▫ Loop detector ⁞ Midpoint		

4.2 Design of Spot Speed Methods for Real-Time Provision ...

shock to travel along the distance "$\Delta x/2$"). On the contrary, the horizon of the "forecasted" situation diminishes with the reduction in the length of the section. In addition, the immediacy in reporting rapidly evolving traffic transitions and the accuracy of the measurement in stationary conditions improves in the case of shorter sections. At the limit (i.e. infinite number of detectors, which is equivalent to vehicle tracking) the system would have no measurement error, but null forecasting capabilities. A trade-off is identified. However it must be recognized that improvements in accuracy and immediacy are on solid ground and necessary conditions, while the improvements of forecasting capabilities are based on more uncertain assumptions. This means that loop detector spacing should be kept short to delimit the measurement error, but also that further reductions will not be translated into more precise real-time information. This assertion provides a conceptual justification to the probably discouraging fact that any accuracy improvement in the current travel time measurement will be swamped out by the evolution of the traffic state until the forecasting horizon.

In spite of this, the relationship between "Δx" and the forecasting horizon is useful for setting the parameters of the real-time information system. The desired forecasting horizon depends on "Δt", the corridor length "$d_C = \sum_{j \in C} \Delta x_j, \forall j \in C$", and the forecasted traffic conditions. It is then possible to obtain suitable relationships between "Δx", "Δt" and "d_C", which approximately match the required real-time forecasting horizon with the Midpoint Algorithm forecasting capabilities, in the most probable traffic states.

Using the time-space diagrams in Fig. 4.1, where it is assumed that congested conditions will eventually cover the whole corridor, and recalling that "Δt" should be kept short, the following relationships can be obtained:

$$\Delta t \leq \left(\frac{v_f - u}{v_f \cdot u}\right) \cdot \frac{\Delta x}{2} \quad \text{from type (b) transitions} \quad (4.3)$$

$$d_C \approx \frac{\Delta x}{2} \cdot \left(1 + \frac{v_f}{u}\right) - v_f \cdot \Delta t \quad \text{from type (a) transitions} \quad (4.4)$$

$$d_C \approx \frac{\Delta x}{2} \cdot \left(1 + \frac{v_{congestion}}{w}\right) - v_{congestion} \cdot \Delta t \quad \text{from type (c) transitions} \quad (4.5)$$

For a given detector spacing "Δx", Eq. 4.3 provides guidance in the selection of the appropriate updating interval, "Δt". This selection is quite flexible, given the range of possible variation in "u" (i.e. 5–15 km/h). Preference must be given to shorter "Δt". Equations 4.4 and 4.5 are useful in order to select the length of the target corridor, as a function of "Δx" and "Δt". In general, Eq. 4.5 provides a lower bound. This should prevail, recalling the shorter the better concept, regarding corridor length.

Equations 4.3–4.5 could be easily modified in case of using whole section constant algorithms. It is only necessary to substitute "$\Delta x/2$" by "Δx".

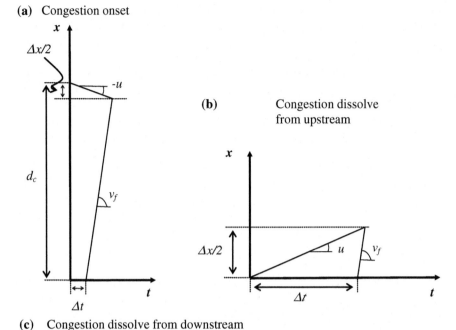

Fig. 4.1 Adequate relationships between corridor and section length

4.2.3 Applications of Other Constant Speed Methods: The Range Method

It can be seen that the immediacy requirement is fulfilled by all standard spot speed travel time methods (e.g. Midpoint, Constant Upstream, Downstream, Conservative or Optimistic). The same cannot be asserted in relation to their forecasting capabilities. Note that the symmetry in relation to the detector location, a property exhibited by the Midpoint Algorithm, allows using the same standard algorithm for all types of transitions. In contrast, the Constant Downstream method would imply forecasting capabilities in case of congestion onset or congestion dissolve from downstream, but not when the congestion dissolves from upstream. The opposite is true for the Constant Upstream method. The forecasting capabilities of these methods, if applied in the correct context, would be "higher" than the Midpoint ones, because the measured speed is extrapolated over a longer distance. The practical implementation of this strategy would be more complex, as it requires the detailed determination of the type of transition on each section.

A better and easier way to deal with the forecasting potential of whole section constant speed methods is by computing the possible range of travel time variation in these sections. Note that the Constant Conservative and the Constant Optimistic

4.2 Design of Spot Speed Methods for Real-Time Provision …

Table 4.3 Determination of traffic conditions on section "j"

	Section "j" traffic conditions		Loop "$j-1$" (upstream end)	Loop "j" (downstream end)
Time evolving conditions	Congestion onset	#1		$v_j^{(t)} + \varepsilon_{v_j}^{(t)} < v_j^{(t-1)} - \varepsilon_{v_j}^{(t-1)}$
	Partial section congestion onset	#2	$v_{j-1}^{(t)} + \varepsilon_{v_{j-1}}^{(t)} < v_{j-1}^{(t-1)} - \varepsilon_{v_{j-1}}^{(t-1)}$	$v_j^{(t)} \pm \varepsilon_{v_j}^{(t)} \cap v_j^{(t-1)} \pm \varepsilon_{v_j}^{(t-1)} \neq \Phi$
	Congestion dissolve	#3		$v_j^{(t)} - \varepsilon_{v_j}^{(t)} > v_j^{(t-1)} + \varepsilon_{v_j}^{(t-1)}$
		#4	$v_{j-1}^{(t)} - \varepsilon_{v_{j-1}}^{(t)} > v_{j-1}^{(t-1)} + \varepsilon_{v_{j-1}}^{(t-1)}$	
	Contradiction	#5	Conditions (#1) and (#4) or (#2 at $j-1$) and (#3 at j) being fulfilled simultaneously	
Time stationary conditions		#6	$\bigcap_{\forall t_i = t-m, t} v_{j-1}^{(t_i)} \pm \varepsilon_{v_{j-1}}^{(t_i)} \neq \Phi$	$\bigcap_{\forall t_i = t-m, t} v_j^{(t_i)} \pm \varepsilon_{v_j}^{(t_i)} \neq \Phi$
Undetermined—previous traffic state prevails		#7	None of conditions (#1)–(#6) are being fulfilled (stationary conditions not assured.)	
Incident detection	Queue ahead	#8	$v_{j-1}^{(t)} - \varepsilon_{v_{j-1}}^{(t)} > v_j^{(t)} + \varepsilon_{v_j}^{(t)}$	
	Leaving congested section	#9	$v_{j-1}^{(t)} + \varepsilon_{v_{j-1}}^{(t)} < v_j^{(t)} - \varepsilon_{v_j}^{(t)}$	

Note (1) Superscript stands for time interval of calculation
(2) "m" stands for a "reasonable" number of time intervals. "m" should be greater than the number of time intervals taken for a shock to travel along the whole section. This is: $m \geq \Delta x/\Delta t \cdot u$
(3) Valid for homogeneous sections (i.e. same fundamental diagram, $q - k$, applies for the whole section)

methods provide upper and lower travel time bounds if only measurement errors are considered. The inaccuracy of spot speed methods can be seen as the travel time range defined by these two estimations, which dramatically increases in transitional sections (see an empirical example in Fig. 4.6). In fact, the increase in this travel time range can be used as an automatic incident detection scheme within each section (see Table 4.3).

The simple dissemination of this range of possible variation may be useful information for the driver. In addition, the real time information system can be enhanced by using the forecasting capabilities of these methods. This is simply achieved by selecting the conservative approach in case of congestion onset within the section, and the optimistic one for any type of congestion dissolve transition. The determination of the direction of the dissolve shock is not necessary in this case. In case of stationary conditions the Midpoint Algorithm stands as the least bad selection. This avoids the possibility of large punctual errors in partially congested sections. This is why the Midpoint Algorithm should also be applied in case of partial congestion growing or in contradictory traffic state determination (see Table 4.3). The added complexity of the transition type determination in this "Range Method" is minimal.

4.3 Spot Speed Methods Improvement in Stationary Conditions

We have shown the potential of spot speed travel time estimation methods in rapidly evolving traffic conditions. However, the accuracy of these methods under stationary conditions remains an issue. Recall from Table 4.1 that this performance is bad (i.e. significant underestimation) in fully congested sections. This error can be high, as it will accumulate across all the sections simultaneously congested, which can be several. The performance will also be very bad in stationary partially congested sections. Although this error can be huge at the sectional level, its contribution to the corridor travel time can be less dramatic, as it is only incurred at the section containing the active bottleneck. It is assumed here that the target corridor is composed of several individual freeway sections.

Two possibilities for improvement are presented: firstly, an intelligent smoothing process that eliminates the volatility of the estimations in stationary periods. This enhances the credibility of the system for users. Secondly, a data fusion scheme to make the most of all possible types of available data.

4.3.1 Intelligent Smoothing Process

In a stationary traffic state, individual travel times are realizations of a random variable resulting from the stochastic nature of driver behavior. Deprived of the time smoothing process that constituted the trajectory reconstruction in "$T_A(C)$" measurements, and taking into account that the time interval "Δt" must be small in order to rapidly report changing conditions, reduce the horizon of the prediction, and not smooth out travel time variations, "$\hat{T}_T(C)$" estimation suffers from fluctuations. This can be clearly seen by realizing that the variance of the time mean speed estimation is reduced with the increase of the number of observations (see Eq. 4.6). Therefore, as small "Δt" implies fewer observations, higher average speed variance results, and this leads to volatile travel time estimations during small time intervals.

Severe travel time fluctuations over consecutive time intervals damage the drivers' perceived credibility of the information system. Therefore, a time smoothing process is necessary. Moving average or exponential smoothing methods have usually been proposed (Cortés et al. 2002; Treiber and Helbing 2002). However, these standard smoothing processes imply a delay in the detection of speed cha nges that is not attributed to fluctuation but to passage of a shock. This implies a loss of the immediacy benefits of "$\hat{T}_T(C)$".

An intelligent smoothing process is proposed here, which smooths out travel time fluctuations while preserving immediacy in the detection of significant speed changes. The method determines whether a speed variation is a random fluctuation and therefore must be smoothed, or whether it is a consequence of a change in the

4.3 Spot Speed Methods Improvement in Stationary Conditions

traffic state and therefore must not be smoothed in order to maintain the immediacy of the information.

Specifically, consider individual vehicular speed as a random variable "V" whose mathematical expectation is "θ" and variance "σ_v^2" over a stationary period. The arithmetic mean of the speed observations within "Δt" is an unbiased estimator of "θ". It can be shown that the variance of the sample mean estimator is given by:

$$Var(\bar{v}) = \frac{\sigma_v^2}{n} \quad (4.6)$$

where "$\bar{v}$" is the sample mean speed and "n" the number of observations within "Δt". Considering the central limit theorem, the sample mean random variable is normally distributed, and therefore the absolute error of the estimation, "ε_v", can be expressed as:

$$\varepsilon_v = (prob.\ level) \cdot \frac{\sigma_v}{\sqrt{n}} \quad (4.7)$$

With a 90 % confidence for the estimation, the probability level takes approximately a value of 2 in the Normal Distribution. Then, the maximum relative error, "e_v", in the estimation of the time mean speed of a stationary traffic stream is:

$$e_v = 2 \frac{C.V._{(v)}}{\sqrt{n}} \quad (4.8)$$

where "$C.V._{(v)}$" stands for the individual vehicle speed coefficient of variation, $C.V._{(v)} = \sigma_v/\theta$. A rough approximation of the speed "$C.V._{(v)}$" suffices, as Eq. 4.8 is steered more strongly by the subjective probability level threshold and the number of observations. Consider for instance the values reported in Soriguera and Robusté (2011) of approximately 0.15. This allows computing the maximum errors on the average speed estimation, in each loop detector and every time period, which can be assumed to be due to stochastic random fluctuations.

A speed variation between consecutive time intervals is considered large, and therefore a state transition, if the intersection of the respective confidence intervals is null. In this case the speed measurement will not be smoothed. Table 4.3 provides more insights in the detection of traffic state transitions from loop measurements. Otherwise, speed variations are considered fluctuations, and smoothed out. The smoothing process consists of a moving average over the last "M" time intervals or since the last transition, whatever occurs first. The "M" smoothing period is obtained by imposing a low probability of obtaining fluctuations in the estimation larger than a particular threshold while in stationary conditions. This issue is addressed in the next section.

4.3.2 Granularity in the Disseminated Information

Given the intrinsic lack of accuracy of the short term predicted travel time, it would make no sense to disseminate very detailed information. In addition, it is the rough estimation that is useful to the driver more than the detail. For instance it would be meaningless to disseminate information with a granularity finer than 1 min, even for very short trips. For longer trips, 5 min granularity may suffice. It could be reasonable to assume a minimum granularity of 1 min in the disseminated information, or the integer number of minutes approaching 10 % of the travel time, whatever is larger. The definition of this granularity can be used to determine a threshold level for the maximum admissible travel time fluctuations in stationary traffic states as all the fluctuations finer than the granularity will be washed out. Note that if the maximum acceptable fluctuations, ε_T, are set equal to the granularity of the dissemination strategy, fluctuations in the disseminated information will not be above two times the granularity. This could be a desirable strategy, and can be formulated as follows:

$$\varepsilon_T = \max\left(g_{\min}, e_T \cdot \hat{T}_T(C)\right) = \max\left(g_{\min}, e_T \cdot \sum_{i=1}^{N} \hat{T}_T(A_i)\right)$$

$$\approx \max\left(g_{\min}, e_T \cdot \frac{\Delta x}{M} \sum_{i=1}^{N} \sum_{j=1}^{M} \frac{1}{\overline{v}_{i,j}}\right) = \max\left(g_{\min}, e_T \cdot \Delta x \cdot N \cdot \hat{E}\left[\frac{1}{\overline{v}_{i,j}}\right]\right)$$

(4.9)

where "$g_{\min}$" is the minimum granularity for short corridors, which usually can be set to 1 min and "e_T" is the longer corridor acceptable granularity in terms of a percentage of the corridor travel time. 10 % can be an adequate selection. In Eq. 4.9 it is assumed that all the "N" detectors belonging to the corridor define "$N-1$" sections of approximately equal length, "Δx", and all of them are in the same stationary traffic state. $\hat{E}\left[\frac{1}{\overline{v}_{i,j}}\right]$ represents an estimation of expected pace (i.e. inverse of speed) computed from the sample mean of average speeds measured during "Δt", "$\overline{v}_{i,j}$", for each one of the "N" detectors (i) and over "M" time periods (j). "M" defines the length of the required smoothing period, which acts as the decision variable.

$$\hat{E}\left[\frac{1}{\overline{v}_{i,j}}\right] = \frac{1}{N \cdot M} \sum_{i=1}^{N} \sum_{j=1}^{M} \frac{1}{\overline{v}_{i,j}}$$

(4.10)

The error in "$\hat{T}_T(C)$" is a result of its variance. And because all the composing "$\hat{T}_T(A_i)$" can be considered independent and identically distributed in a unique stationary traffic state:

4.3 Spot Speed Methods Improvement in Stationary Conditions

$$Var(\hat{T}_T(C)) = Var\left[\sum_{i=1}^{N}\hat{T}_T(A_i)\right] = Var\left[\frac{\Delta x}{M}\sum_{i=1}^{N}\sum_{j=1}^{M}\frac{1}{\overline{v}_{i,j}}\right] = \frac{\Delta x^2 \cdot N}{M}Var\left[\frac{1}{\overline{v}_{i,j}}\right] \quad (4.11)$$

Then:

$$\varepsilon_T = \sqrt{Var(\hat{T}_T(C))} = \sqrt{\frac{N}{M}} \cdot \Delta x \cdot C.V._{\left(\frac{1}{\overline{v}_{i,j}}\right)} \cdot \hat{E}\left[\frac{1}{\overline{v}_{i,j}}\right] \quad (4.12)$$

Using Eqs. 4.9 and 4.12, given the Normal distribution of the sample mean:

$$M \geq \begin{cases} \dfrac{(prob.level)\cdot N\cdot \Delta x\cdot C.V._{\left(\frac{1}{\overline{v}_{i,j}}\right)} \cdot \hat{E}\left[\frac{1}{\overline{v}_{i,j}}\right]}{g_{\min}} & \text{if } e_T \cdot N \cdot \Delta x \cdot \hat{E}\left[\frac{1}{\overline{v}_{i,j}}\right] \leq \varepsilon_T \\[2ex] \dfrac{(prob.level)\cdot C.V.^2_{\left(\frac{1}{\overline{v}_{i,j}}\right)}}{e_T^2 \cdot N} & \text{if } e_T \cdot N \cdot \Delta x \cdot \hat{E}\left[\frac{1}{\overline{v}_{i,j}}\right] > \varepsilon_T \end{cases} \quad (4.13)$$

where the coefficient of variation of the average paces measured over "Δt" can be computed from a pre-sample. As an order of magnitude, and using the data presented in Sect. 4.4, where "$\Delta t = 3$ min", "$C.V._{\left(\frac{1}{\overline{v}_{i,j}}\right)}$" takes values of 0.035 in case of free flowing traffic conditions and 0.23 for congested ones. Considering typical error values and congested conditions (i.e. worst case scenario), Eq. 4.13 can be approximated by:

$$M \geq \frac{10}{N} \quad (4.14)$$

4.3.3 Stationary Conditions: Opportunities for Data Fusion Schemes

Some situations may exist where both, "$\hat{T}_T(C)$" from loop detectors and "$T_A(C)$" by means of some direct measurement technology, are available. In this case, the real-time travel time information system could be enhanced by using the potentialities of each type of measurement (i.e. the immediacy and forecasting capabilities of spot speed algorithms in evolving conditions and the accuracy of direct measurements in stationary conditions). If stationary conditions in the loop measurements (see Table 4.3) and in "$T_A(C)$" (i.e. no large variation, as defined in the Sect. 3.1, has occurred in the last "$m*$" time intervals) are measured, "$T_A(C)$" can be disseminated by the real-time information system. Otherwise, spot speed methods must prevail. Note that the required stationary in "$T_A(C)$" measurements implies that "$m*$" can be smaller than "m" (because there is no uncertainty of what is happening in between detectors).

This data fusion scheme avoids the measurement errors of spot speed methods in stationary conditions (i.e. fully congested or stationary partially congested sections). In addition, in the uncommon case of "$N = 2$" (i.e. single section corridors) it is also possible to locate the transition position within the section:

$$T_A(C) = \frac{d_j}{\bar{v}_j} + \frac{\Delta x - d_j}{\bar{v}_{j-1}} \qquad (4.15)$$

$$d_j = \frac{T_A(C) \cdot \bar{v}_j \cdot \bar{v}_{j-1} - \bar{v}_j \cdot \Delta x}{\bar{v}_{j-1} - \bar{v}_j} \qquad (4.16)$$

where "$\bar{v}_j$" and "$\bar{v}_{j-1}$" are the downstream and the upstream average speeds respectively, and "d_j" is the transition distance to the downstream detector.

4.4 Test Site and Empirical Data

The AP-7 turnpike, on the northeastern stretch of the Spanish Mediterranean coast, represents a privileged test site. "$T_A(C)$" are available between any pair of junctions from the tolling system (see Chap. 5) "$\hat{T}_T(C)$" can be obtained from the installed double loop detectors. All data are obtained as a 3-min average (i.e. "Δt" = 3 min). The test corridor, between the "Cardedeu" junction and the toll plaza at "La Roca del Vallès" (see Fig. 4.2), consists of 3.62 km on the southbound direction towards

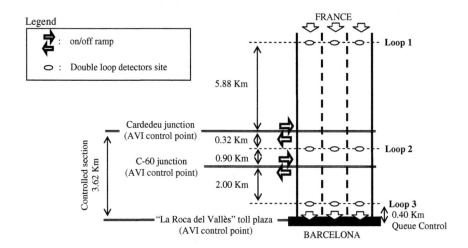

Fig. 4.2 Test site layout

4.4 Test Site and Empirical Data

Barcelona. In this example, the target corridor is defined by Loops 2 and 3 (i.e. "N" = 2) and it can be accepted as a single section, "Δx" = 2.9 km. The length of this section would have traditionally been considered too large for the application of spot speed travel time estimation methods, because of possible measurement errors in transition conditions as large as 10 min (relative error ~ 100 %) (see Eq. 4.1). However it will be shown here that it still provides significant benefits in case of real-time information systems.

Note that despite the large "Δx", the parameters of the real-time information system can be considered adequate, as Eqs. 4.3–4.5 are approximately fulfilled.

Test data were obtained on Thursday June 21st, 2007, a very conflictive day in terms of traffic in the selected stretch. Problems started around 12:39 when a strict roadblock at "La Roca del Vallès" toll plaza was set up by the highway patrol, reducing its capacity and consequently the output flow. This, in addition to the high traffic demand of the turnpike at that time, caused severe queues to grow rapidly. In view of this fact, and approximately 45 min after the setting of the roadblock, when the queue was already spanning around 5 km, the service rate of the roadblock was increased. This reduced the queue growing rate. Until 14:33 when the patrol roadblock was removed, queues had not started to dissipate. In addition, at 14:51 when queues were still dissipating, the breakdown of a heavy truck just downstream of Loop 2 blocked one of the three lanes. In turn, this caused queues to start growing upstream again. Finally, when the broken-down truck was removed at 16:06, the queues started to dissipate again, flowing at capacity. The high and unanticipated demand at "La Roca" toll plaza, as a consequence of the queue discharge, exceeded the capacity of the toll gates, causing small queues to grow at this location between 16:06 and 16:30. A complete sketch of traffic evolution on the test site for this particular day can be seen in Fig. 4.3, drawn at a scale to match empirical data presented in Figs. 4.4 and 4.5. From Fig. 4.4 it can also be seen how Loop 3 is impacted by the slowing for the toll plaza in free flowing conditions. The average "v_f" on the corridor can be set to 110 km/h.

This particular congestion episode on the test site is illustrative, as different types of transitions and stationary periods arise. Specifically:

- Free flowing traffic most of the day.
- Congestion onset (i.e. type (*a*) transition) between approximately 12:33 and 13:21, when the shocks were travelling upstream at an approximated speed of 8.3 km/h. Note from Fig. 4.4 that approximately 21 min were necessary for the transitions to travel from Loop 3 to Loop 2, the 2.9 km section length.
- Congested traffic on the whole corridor between 13:24 and 14:30.
- Congestion dissolve between 14:33 and 15:00, when the queue was dissolving from downstream (i.e. type (*c*) transition) due to the removal of the patrol roadblock, at an approximated speed of 14.5 km/h. Note from Fig. 4.4 that it took up to 12 min for the queue to dissolve between Loop 3 and Loop 2.
- Partially stationary congested corridor due to a lane closure (resulting from the truck breakdown) nearby the upstream end of the section between 15:12 and 16:06.

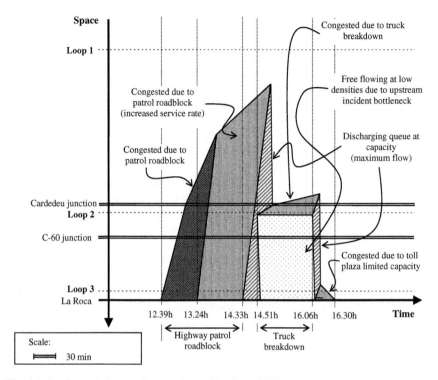

Fig. 4.3 Traffic evolution on the test site on 21st June, 2007

4.5 Empirical Results

Figure 4.5 shows the travel times that would have been disseminated by the real-time information system in case of using AVI arrival based travel times (i.e. "$T_A(C')$") or alternatively, by using the Midpoint Algorithm estimated travel times (i.e. "$\hat{T}_T(C)$"). The desired information, "$T_D(C')$" (AVI departure based) is also plotted. Recall that this last information would not have been available in the real-time operation of the system.

Information delay of arrival based travel times in transition conditions is evident. The benefits of using the Midpoint algorithm, resulting from its immediacy and forecasting capabilities, are also clear. The performance of the Midpoint Algorithm in time stationary conditions is, as expected, good in free flowing conditions, while systematic travel time underestimation is observed when congestion spans the whole stretch. In case of a stationary partially congested section, totally unrealistic travel times are obtained. Note that in this last situation, between 15:12 and 16:06, the queue only spans a few upstream meters of the section, but affects Loop 2. Travel times are similar to the free flowing situation, but very low speeds are measured at the loop location, which completely bias the Midpoint estimation.

4.5 Empirical Results

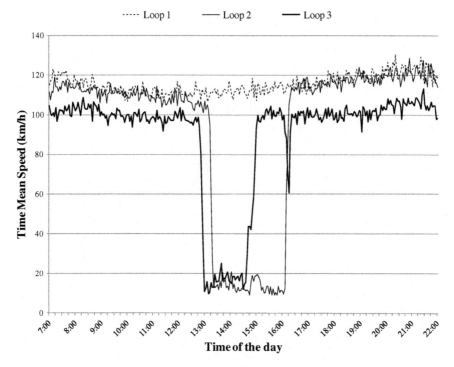

Fig. 4.4 Time mean speed (3 min average) for the whole section measured at loop detector sites

Figure 4.5 also shows the results of applying the described smoothing process for stationary conditions (with "M" = 5). This process can help in avoiding the detrimental fluctuations of disseminated information, but its contribution to the performance of the system is marginal.

Figure 4.6 is aimed to show the actual disseminated information, given a specified travel time granularity (e.g. 1 min). It also shows how simple computation of the measurement error range allows adding the concept of reliability to the estimation. It can be seen how this range dramatically increases when transition conditions within the section appear. The determination of the type of transition (i.e. congestion onset, dissolve or stationary) allows informing accordingly (e.g. by using upstream dynamic message signs), and enhancing the real-time information system by using the "range method". This is only a smart combination of Midpoint, Constant Conservative and Constant Optimistic estimations.

In Fig. 4.6 it can be seen that the range method makes the most of the immediacy and forecasting capabilities of spot speed methods. However, inaccuracy in stationary traffic states remains an issue. This is addressed in Fig. 4.7 where the results of the proposed data fusion scheme are plotted. Of course this approach requires both direct travel time measurements and indirect estimations being available.

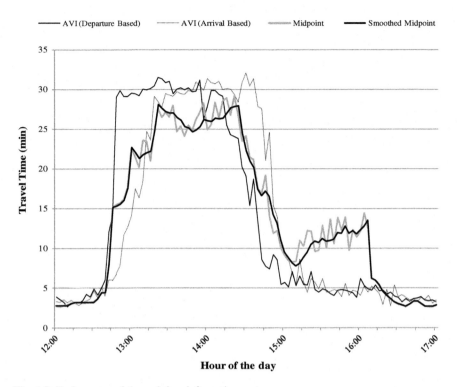

Fig. 4.5 Performance of the real-time information system

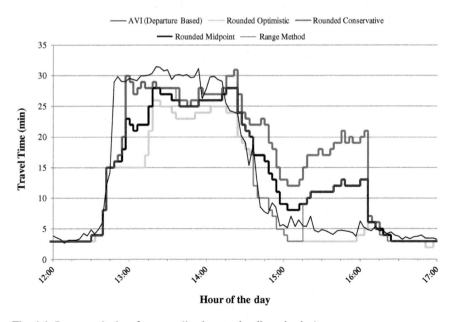

Fig. 4.6 Range method performance (1 min granular dissemination)

4.5 Empirical Results

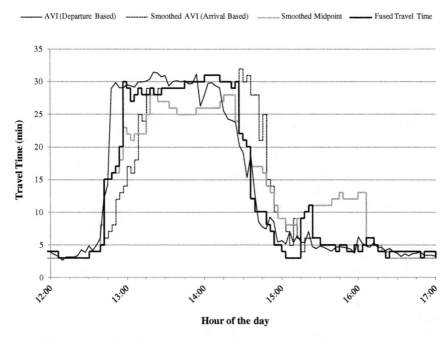

Fig. 4.7 Data fusion scheme performance (1 min granular dissemination)

The lack of accuracy of the system at the start of the stationary transition is seen in Fig. 4.7. This behavior should not be unexpected, as during "m^*" = 3 time intervals, the system detects a congestion onset on a partially congested section, so that the Midpoint Algorithm is selected. It is not until after "m^*" time intervals that the system is able to assume stationary conditions. It is also at this point in time when the data fusion scheme allows locating the inner section incident, by using Eq. 4.16. For the present application "$d_j \approx 2.9$". This confirms that the incident is approximately located at the upstream loop detector (Loop 2) location. In any case, the performance of the real-time information system in this stage is at least acceptable.

4.6 Conclusions

In a real-time context, the main advantage of travel times estimated from loop measurements is the possibility of obtaining true travel times; this is to obtain a virtual measurement for a vehicle travel time before the end of its journey. This provides benefits in the immediacy of the detection of transitions in the traffic stream state. However, these benefits are obscured by the lack of accuracy of these measurements. It has been shown in the chapter how an adequate selection of the speed extrapolation method to the freeway section allows treating inaccuracies in a way that is transformed into forecasting capabilities. The overall performance of these methods in real-time is better than others that only focus on the accuracy of the measurement.

From the chapter it can be concluded that a real-time freeway travel time information system can be supported by simply using the traditional Midpoint Algorithm for travel time estimation using the speed measurements at loop detectors. It is only needed to select appropriate parameters for the system. This means adequate target corridor lengths for the disseminated information, adequate updating intervals, and adequate detector spacing. Guidelines for these selections are provided in the chapter. In particular, it is shown that detector spacing can be on the order of a small number of kilometers without greatly penalizing the system. The reader must be warned that most of these guidelines are rough approximations that hold in the most common and general traffic situations. However, particular situations may exist where they fail. This does not invalidate the general usefulness of the proposed relationships.

Improvements to the Midpoint Algorithm are also developed. Firstly, a method to smooth the fluctuations in stationary periods while preserving the immediacy in reporting traffic transitions is proposed. Secondly the "range method" is constructed, using different constant speed extrapolation methods as building blocks. This method can be used as an incident detection scheme and also to enhance the forecasting capabilities of spot speed based methods in transition conditions.

Finally, it has been stated in the chapter that directly measuring travel times provides accuracy benefits, though delayed information. In stationary conditions the delay in the information is not an issue, even in the case of a real-time system. Therefore, if available, this accurate information must be used in these conditions. A data fusion scheme capable of taking the better of each type of measurement is proposed.

The contribution of this chapter should be considered as a simplistic but possible solution to the short-term freeway travel time forecasting problem. The main property of this solution is that is practice ready (i.e. easy, cheap and based on currently available data). However, this should not be a disincentive for further research on a more elaborate and accurate solution, which may become a feasible option to practitioners in the near future.

References

Cortés, C. E., Lavanya, R., Oh, J., & Jayakrishnan, R. (2002). General-purpose methodology for estimating link travel time with multiple-point detection of traffic. *Transportation Research Record, 1802*, 181–189.

Daganzo, C. F. (1990). Fundamentals of transportation and traffic operations. Upper Saddle River: Prentice-Hall.

OECD/JTRC. (2010). Improving reliability on surface transport networks. Paris: OECD publishing. ISBN 978-92-82-10241-1.

References

Palen, J. (1997). The need for surveillance in intelligent transportation systems. *Intellimotion, 6*(1), 1–3, 10. (University of California PATH, Berkeley, CA).

Soriguera, F., & Robusté, F. (2011). Estimation of traffic stream space-mean speed from time aggregations of double loop detector data. *Transp Res Part C, 19*(1), 115–129.

Soriguera, F., Rosas, D. (2012). Deriving traffic demand patterns from historical data. In *Proceedings of the 91st Transportation Research Board Annual Meeting*. Washington, D.C.

Treiber, M. & Helbing, D. (2002). Reconstructing the spatio-temporal traffic dynamics from stationary detector data. *Cooper@tive Tr@nsport@tion Dyn@mics 1*, 3.1–3.24.

Chapter 5
Highway Travel Time Measurement from Toll Ticket Data

Abstract Travel time for a road trip is a drivers' most appreciated traffic information. Measuring travel times on a real time basis is also a perfect indicator of the level of service in a road link, and therefore is a useful measurement for traffic managers in order to improve traffic operations on the network. In conclusion, accurate travel time measurement is one of the key factors in traffic management systems. This chapter presents a new approach for measuring travel times on closed toll highways using the existing surveillance infrastructure. In a closed toll system, where toll plazas are located on the on/off ramps and each vehicle is charged a particular fee depending on its origin and destination, the data used for toll collection can also be valuable for measuring mainline travel times on the highway. The proposed method allows estimating mainline travel times on single sections of highway (defined as a section between two neighboring ramps) using itineraries covering different origin–destinations. The method provides trip time estimations without investing in any kind of infrastructure or technology. This overcomes some of the limitations of other methods, like the information delay and the excess in the travel time estimation due to the accumulation of exit times (i.e. the time required to travel along the exit link plus the time required to pay the fee at the toll gate). The results obtained in a pilot test on the AP-7 toll highway, near Barcelona in Spain, show that the developed methodology is sound.

Keywords Highway travel time measurement · Toll highways · Toll ticket data

5.1 Introduction

There is common agreement among drivers, transportation researchers and highway administrations that travel time is the most useful information to support trip decisions (users) and to assess the operational management of the network (administrators), (Palen 1997).

In response to these needs for accurate road travel time information, researchers and practitioners from all over the world have worked hard in this direction. During the last two decades, research efforts have been focused on the indirect estimation of road travel times, using the fundamental traffic variables, primarily each vehicle's speed observed at discrete points in the freeway. The prominence of this approach results from the fact that, for ages, these have been the unique available traffic data, as provided by inductance loop detectors. Advances in this research area have been huge as it demonstrates a vast related literature. The efforts made in improving the accuracy of speed estimations from single loop detectors should be emphasized (Coifman 2001; Dailey 1999; Mikhalkin et al. 1972; Pushkar et al. 1994). However, although accurate spot speed estimations have been obtained (as in the case of using double loop detectors), travel time estimates could still be flawed due to extrapolating spot measurements to a highway section, with the possibility of different traffic conditions (congested or not) along its length. Note that this problem is greater on highways with a low density of detection sites. One detector site every half kilometer of highway is desirable to reduce the effect of this problem (Hopkin et al. 2001). Several approaches have been published trying to overcome this limitation without falling into the enormous cost of intensive loop surveillance, proposing, for example, different methods for the reconstruction of vehicle trajectories between loop detectors (Coifman 2002; Cortés et al. 2002; Li et al. 2006; van Lint and van der Zijpp 2003). In addition to these problems, it must also be taken into account that the loop speed estimates in the case of stop and go traffic situations do not adequately represent the space mean speed of the traffic stream.

A different approach to the indirect estimation of link travel times using loop detectors consists of comparing the cumulative counts (N-curves) from consecutive loop detectors (instead of using the spot speed measurement at the detector site). In the case of all on and off ramps being monitored, the flow conservation equation can be applied to obtain the travel time in the stretch (Nam and Drew 1996; van Arem 1997). This method does not suffer from previous speed estimation limitations, but must account for loop detector drift that can jeopardize the accuracy of the results.

Lately, research on travel time estimation using loop detector data has focused on direct measurement, consisting of measuring the time interval that a particular vehicle takes to travel from one point to another. To achieve this goal several authors propose a smart use of loop detector data, on the basis of the re-identification of particular vehicles in consecutive loop detectors by means of characteristic length (Coifman and Ergueta 2003; Coifman and Krishnamurthya 2007) or particular inductive signature on the detector (Abdulhai and Tabib 2003; Sun et al. 1998, 1999). An extended approach of these algorithms is the re-identification of particular features of vehicle platoons instead of individual vehicles (Coifman and Cassidy 2002; Lucas et al. 2004). All these last contributions can, however, not be put into practice with the current common hardware and/or software loop configurations. Most operating highway agencies would have to upgrade their systems in the field to accomplish these objectives.

5.1 Introduction

In another order of events, the deployment of ITS (Intelligent Transportation Systems) during the last decade has brought the opportunity of using more suitable traffic data to directly measure travel times (Turner et al. 1998). This is the case with AVI (Automated Vehicle Identification) data obtained, for instance, from the readings of vehicle toll tags or from video license plate recognition. By matching the vehicle ID at different locations on the highway, link travel times can be directly obtained if the clocks at each location are properly synchronized. Another approach for directly measuring travel times is to use automatic floating car data obtained from different technologies such as GPS (Global Positioning Systems) or the emerging cellular phone geo-location. Take as an example the Mobile Century field experiment performed recently in a Californian highway (Herrera et al. 2010). In these schemes travel times are obtained by the real time tracking of probe vehicles, being their number critical for the accuracy of the measurements. Results obtained by Herrera et al. (2010) suggest that a 2–3 % penetration of GPS-enabled cell phones in the drivers' population is enough to provide accurate measurements of the velocity of the traffic stream.

The present chapter focuses on the direct highway travel time measurement using AVI data from toll collection systems. Although this concept is not new (Davies et al. 1989), the contributions found in the literature primarily deal with the usage of ETC (Electronic Toll Collection) data to measure travel times. These systems identify the vehicles by means of on-vehicle electronic tags and roadside antennas located, sometimes ad hoc, on the main highway trunk. Under this configuration the basic problems are the level of market penetration of the electronic toll tags and how to deal with time periods when only small samples are available in order to obtain a continuous measurement of travel times (Dion and Rakha 2006; SwRI 1998). Surprisingly, very few contributions are found related to travel measurement using the primitive configuration of a closed toll system. The concept of a "closed" toll system refers to the fact that the toll a particular driver pays varies depending on the origin and destination of his trip and is approximately proportional to the distance traveled on the highway. In contrast, one has to bear in mind the "open" toll systems, where toll plazas are strategically located so that all drivers pay the same average fee at the toll booth.

This chapter deals with travel time measurement in the typical closed toll system configuration, widely extended in Europe and Japan for a long time, and by the authors knowledge only discussed in Ohba et al. (1999) in the particular case of main highway trunk toll plazas and a single origin destination pair. Under the closed toll configuration, each vehicle entering the highway receives a ticket (traditionally a card with magnetic band or more recently a virtual ticket using an ETC device), which is collected at the exit. The ticket includes the entry point, and the exact time of entry. By cross-checking entry and exit data, the precise time taken by the vehicle to travel along the itinerary (route) can be obtained (obviously clocks at the entry and exit toll plazas are considered to be synchronized). Averages can be obtained from the measurements for all the vehicles traveling along the same itinerary in the network. In relation to using only ETC based travel time estimation, a particular advantage of this configuration is the huge amount of data, since all vehicles have their entry/exit ticket (real or virtual), solving the problem of the market penetration of the ETC devices. However, other problems arise from this configuration, which are discussed in the next section of the chapter. The proposed solution involves the estimation of the single section travel time (i.e. the time required to travel between two consecutive ramps on the highway) and also the exit time for each ramp (i.e. the time required to travel along the exit link plus the time required to pay the fee at the toll gate). Combining both estimations makes it possible to calculate all the required route travel times.

This chapter is organized as follows: first, in Sect. 5.2, the context of the problem and the solution approach are described. In Sect. 5.3, the concept of the algorithm and the basic notation and formulation are provided, keeping the data filtering process and the more mathematical expressions of the algorithm at the end of the chapter in Appendices 5A, 5B and 5C. Section 5.4 presents some modifications of the algorithm for its implementation in real time or off-line configurations. Then in Sect. 5.5, results of the application of the model to the AP-7 highway in Spain are presented. Finally, general conclusions and issues for further research are discussed.

5.2 Objective of the Proposed Algorithm

As stated above, travel time can be obtained by directly measuring the time taken for vehicles to travel between two points on the network, and this seems particularly easy in closed toll highways, where the data needed for the toll collection makes it possible to obtain itinerary travel times for all origin–destination relations on the highway. Despite this apparent simplicity, several problems arise.

In this configuration, travel time data is obtained once the vehicle has left the highway. All direct travel time measurements and also some indirect estimation algorithms (e.g. some application of the N-curves method), suffer from this drawback. This type of travel time measurement, which will be named measured travel time (MTT), represents a measurement of a past situation and involves a great delay in information in the case of long itineraries or congested situations. Another limitation of this data is that travel times are only valid for a particular origin–destination itinerary such that partial on route measurements cannot be obtained.

In order to reduce delay in travel time information (and still directly measure), it is necessary to estimate MTTs for itineraries as short as possible. In a closed toll highway context, this leads to measuring single section travel times (i.e. between consecutive junctions). By doing so, information delay is reduced to a single section travel time. Single section measurements also overcome the limitation of itinerary specific travel times, since travel time estimation for long trips (i.e. more than one single section) can be obtained by adding the different single section travel times that configure the route. This procedure provides valid information for all drivers who pass through the highway section (regardless of whether they have the same origin–destination itinerary or not), and could also enable incident detection applications by tracking down the conflictive highway sections.

5.2 Objective of the Proposed Algorithm

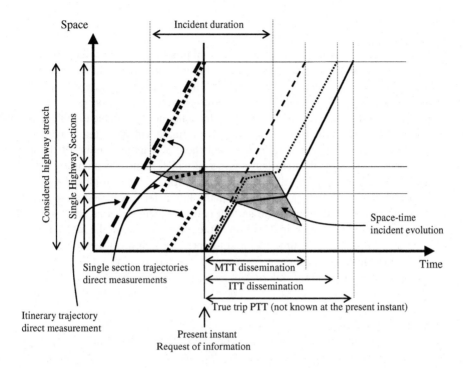

Fig. 5.1 Travel time definitions and their possible implications in the dissemination of information

The itinerary travel time resulting from the addition of the travel times spent in the single sections that form the route at the same instant will be named instantaneous trip travel time (ITT) and assumes that traffic conditions will remain constant in each section until the next travel time update. This estimation is a better approach to the predicted travel time (PTT), which represents an estimation of the expected travel time for a driver entering the highway at the present instant, than MTT. Note that the ITT is a virtual measure in the sense that in fact no driver has followed the trajectory from which this travel time comes.

By way of illustration, Fig. 5.1 shows an example of the implications of different trip travel time constructions. The information delay in the case of trip MTT involves very negative effects in case of dramatic changes in traffic conditions during this time lag (e.g. an incident occurs). The construction of trip ITT by means of single section travel times reduces this information delay and the resulting travel time inaccuracies. As the traffic conditions do not remain constant until the next single section travel time update, the trip ITT also differs from the true PTT. Note that the intention in Fig. 5.1 is to show the maximum differences that could arise between MTT, ITT and PTT. This happens in case of rapidly evolving traffic conditions, for instance when an incident happens. Due to the deliberate construction of Fig. 5.1, MTT misses the onset of congestion, while the ITT is able to detect it. This is the case where benefits of using ITT as opposed to MTT would be maximized.

In this context, the main goal of the algorithm proposed in the present chapter is to obtain the required single section travel times from the available closed toll system data with no additional surveillance infrastructure. Obviously, a naïve method could be to only consider measurements between consecutive entry and exit ramps. This solution may reduce excessively the amount of available data in certain sections of the network, where the volume of traffic entering and leaving the highway at consecutive junctions is low, but there is a large volume of through traffic. Even in the case of interurban highways (which is the common case where highways are equipped with toll booths at each entrance and exit-closed highway systems) where consecutive junctions are many kilometers away and a significant number of drivers traveling a single section could be achieved, this naïve method does not account for the "exit time" (i.e. the time required to leave the highway) and the "entrance time" (i.e. the time required to enter the highway). It must be taken into account that the measurement points are located at the very end of the on/off ramps, sometimes a couple of kilometers away from the main highway trunk (see Fig. 5.2). The exit time includes the time required to travel along the exit ramp (deceleration and overcoming the distance along the ramp until reaching the toll booth) plus the time required to pay the fee (perhaps with a small queue). In this situation, if the time to travel along a particular route, composed of several single sections, is calculated by simply adding the single itinerary travel times, the resulting travel time would be largely overestimated, because it would include as many exit and entrance times as there are single sections comprising the itinerary.

Another solution to estimate single section travel times that would overcome the previous problems, is to install roadside beacons on the main highway trunk to detect vehicles equipped with an ETC system tag, and convert the traditional closed system into an ETC based travel time measurement system. However, and in addition to the high implementation costs (up to $100 000 per lane and measuring

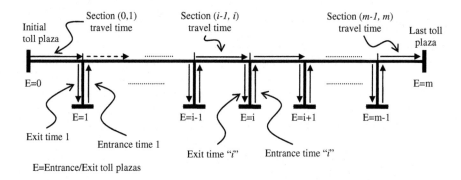

Fig. 5.2 Closed highway network travel time elements

point if the overhead gantry is not available), the market penetration of the toll tags become a problem, as referenced in the introduction.

The algorithm presented here estimates the single section travel times without reducing excessively the amount of available data, and makes it possible to split this time into the main highway trunk travel time and the exit time, without any additional surveillance equipment. The exit time is a very useful collateral result for highway operators, as it is an indicator of the toll plazas' level of service.

5.3 Estimation of Single Section Travel Times: The Simple Algorithm's Underlying Concept

For each particular vehicle "k" running along a highway with a closed tolling system, the travel time spent on its itinerary between origin "i" and destination "j", expressed as "$t_{i,j,k}$", can be obtained by matching the entry and exit information recorded on its toll ticket. As the toll ticket also includes the type of vehicle, only the observations corresponding to cars (including also motorbikes) are considered. Trucks are not considered, due to their lower speeds, which would bias the travel time estimations in free flow conditions. Of course, truck observations could be considered, as a family apart, if one is interested in obtaining specific free flow travel time for trucks. Note that the objective is to provide accurate travel time information for the driver of a car whose travel speed is considered as safe and comfortable under the existing traffic conditions by the average driving behavior.

The travel time information updating interval is defined as "Δt". Then, "$t_{i,j}^{(p)}$" refers to a representative average of the "$t_{i,j,k}^{(p)}$" data obtained in the "p" time interval (i.e. all the "k" vehicles that have exited the highway at "j", coming from "i" between the time instants "$p - \Delta t$" and "p"). To obtain this representative average "$t_{i,j}^{(p)}$" is not an easy task; in fact this is the key thing in the only ETC based travel time estimation systems. Problems arise from the nature of these data, with high variability in the number of observations (depending on the selected itinerary and time period) and different types of outliers to be removed to avoid producing erroneous travel time estimates. The range of solutions is huge, from the simple arithmetic mean, to the complex data-filtering algorithms developed by Dion and Rakha (2006), where a good overview of these filtering methods and current applications is also presented. The data-filtering algorithm used in the present approach is presented in Appendix 5A.

Once representative averages of travel times in all itineraries for the time interval just elapsed are obtained, the next step is to calculate the single section travel time and the exit time.

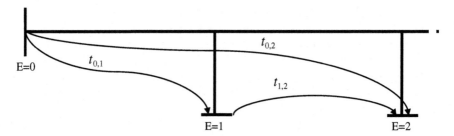

Fig. 5.3 Section (0,1) travel time estimation

5.3.1 Basic Algorithm

Consider the highway stretch between entrance 0 and exit 1. The average travel time in this single itinerary, "$t_{0,1}$"[1] can be divided into two parts: the single section travel time "$t_{s(0,1)}$" and the exit time "$t_{ex(1)}$" (see Fig. 5.2).

$$t_{0,1} = t_{s(0,1)} + t_{ex(1)} \tag{5.1}$$

By subtracting different travel times of selected itineraries, the single section travel times and the exit times can be obtained by canceling out the exit times[2] (see Fig. 5.3). Then for the (0,1) itinerary:

$$t_{s(0,1)} \approx t_{0,2} - (t_{1,2} - t_{en(1)}) \approx t_{0,3} - (t_{1,3} - t_{en(1)}) \approx \cdots \tag{5.2}$$

$$t_{ex(1)} = t_{0,1} - t_{s(0,1)} \tag{5.3}$$

where, "$t_{en(1)}$" is the entrance time at on-ramp 1 (i.e. the time required to travel along the entrance link).

In a general expression for itineraries with an entrance different from the initial toll plaza, located on the main trunk of the highway, Eq. 5.1 should be rewritten as:

$$t_{i,i+1} = t_{en(i)} + t_{s(i,i+1)} + t_{ex(i+1)} \quad \forall i = 1, \ldots, m-1 \tag{5.4}$$

where, "m" is the last toll plaza on the highway, located on the main highway trunk.

---

[1] Note that to simplify the notation and clarify the concepts, the superscript "(p)" is omitted in Sects. 5.3.1 and 5.3.2. The original notation will be recovered in Sect. 5.4.1, where "(p)" plays an important role.

[2] Equations (5.1), (5.3), (5.4), (5.6), (5.7), (5.9), (5.11), (5.21), and (5.29) are not exactly true. The lack of accuracy can be seen as a notation simplification that helps to clarify the concept, and is further detailed in Appendix 5B.

5.3 Estimation of Single Section Travel Times ...

The entrance time "$t_{en(i)}$" can be estimated as a constant parameter for each entrance "i". It is calculated by considering a constant acceleration from the toll booth (where the vehicle is stopped) until the end of the entrance ramp (where it can be considered the vehicle is traveling at 90 km/h). The length of the entrance ramps are designed so that this speed could be achieved with typical vehicle acceleration rates. In addition, a reaction time of 5 s at the toll booth is added. Note the implicit assumption related to this constant entrance time: free access to the highway. There isn't any type of ramp metering scheme, neither any queue to join the main trunk traffic. This assumption generally holds in closed toll highways, because the ticketing at the toll booth smooth the entrance rate to the ramp, which, in general, is lower than the merging rate when entering the main highway traffic stream. However, in high demanded on-ramps and heavy congested highway main trunk, on-ramp queues should not be discarded, even though some of the queue is shifted upstream of the toll booth. Therefore constant entrance times should be considered as a simplification of the method that must be taken into account before each particular application.

The solution to this limitation is simple: to measure or to estimate entrance times, between the toll booth and the effective incorporation to the mainline. In order to measure it some type of additional surveillance equipment at the incorporation is needed (e.g. a vehicle identification control point—ETC antenna, automatic license plate reading—or a loop detector). Then entrance times could be directly measured or indirectly estimated using one of the multiple methods described in the introduction of the chapter (e.g. from difference between cumulative counts). If this additional surveillance equipment is not available, and as the objective is to provide travel time estimations without investing in any kind of infrastructure or technology (as long as the instant that each vehicle passes through the toll plaza is registered), an alternative solution could be proposed if entrance time variations are considered to be critical. This consists in the estimation of a maximum merging rate and to compare it with the entering flow at the on-ramp toll booth. Using deterministic queuing diagrams, the delay at the on-ramp could be easily obtained. The main problem here could be the estimation of the maximum merging rate, which depends on the traffic conditions on the main highway trunk. However, it has been experimentally found (Cassidy and Ahn 2005) that the maximum rate at which vehicles can enter a congested freeway from an on-ramp is a fixed proportion of the downstream freeway flow.

Following the same logic as in Eq. (5.2), the general expression to calculate the single section travel time and the exit time can be formulated as:

$$t_{s(i,i+1)} \approx (t_{i,i+2} - t_{en(i)}) - (t_{i+1,i+2} - t_{en(i+1)})$$
$$\approx (t_{i,i+3} - t_{en(i)}) - (t_{i+1,i+3} - t_{en(i+1)}) \approx \cdots \quad (5.5)$$

$$t_{ex(i+1)} = t_{i,i+1} - t_{s(i,i+1)} - t_{en(i)} \quad (5.6)$$

Equations (5.5) and (5.6) involve estimating single section travel times using different origin–destination itineraries. Note that in some traffic situations the use of different lanes is not independent of the destination. Therefore, if there is a significant speed gradient between these lanes, this implies that travel times in the highway section are not identical independent of the vehicle destination. This could lead to a significant error.

However these situations only significantly arise near off-ramps, for instance when the demand for exiting the tollway exceeds the capacity of the off-ramp. This results in a spill-back of the queue into the main highway trunk, congesting the rightmost lanes and reducing the capacity of the rest of the lanes due to the "friction" between congested and not congested lanes, and due to last minute lane changes. The situation would result in the rightmost lanes congested, and only composed of vehicles wishing to exit in the next off-ramp, while on the other lanes traffic stream would be uninterrupted, but not at free flow speed, and composed by drivers heading to all other destinations. Therefore this section travel times depend on the destination, but only on two groups of destinations, the next off-ramp and all the others. Empirical evidence shows (Muñoz and Daganzo 2002) that after some few kilometers, even a wide multilane highway becomes FIFO. Therefore the lane effect only arises for a limited length. In case of interurban highways, which may be narrower and the sections between junctions longer, it is even more appropriate to assume that multi-pipe traffic states (i.e. non-FIFO congested regimes) will be confined within the single section defined by the off-ramp.

Then note that this does not imply any drawback to the proposed algorithm since the implicit assumption is that all vehicles traversing a freeway section heading to all destinations except the next one have similar travel times in the section. This results from the fact that in order to calculate a single section travel time, we operate with itineraries heading to the same destination (to cancel out exit times); therefore, this last section destination specific travel time is also cancelled out.

The only implication of this destination specific section travel time, being this section the last section of the vehicle itinerary, appears in the calculation of the exit time, when last sections of the itinerary are considered. The traffic situation exemplified above would result in an increased exit time for the off-ramp. This only implies that the exit time does not only take place in the off-ramp, but also queuing in the rightmost lanes of the main highway trunk in the last section.

The traffic situation described could also go the other way around: congestion in the main trunk while less demand than capacity for the off ramp. This would not result in significant non FIFO queues due to speed gradient across lanes in the section, since in common freeway configurations with a constant number of lanes, the main trunk congestion would block the off-ramp until the vehicles reach the exit point. However there exist some specific freeway configurations where an auxiliary lane is available only for the next off-ramp, preventing some vehicles to queue if they do not want to cross the bottleneck, and therefore reducing the extent of the queue. This would result in an uninterrupted flow for vehicles heading to the next exit, while congested in the mainline for all other destinations. The proposed algorithm in this case computes the congested travel time for the section, and a low or eventually even negative exit time for the off-ramp. This exit time would correspond to the sum of two effects: the true exit time and a (negative) time representing the time savings in the last freeway section in comparison to those drivers in the section not heading to the off-ramp. In conclusion, if one wants to disseminate the travel time to that specific off-ramp, by adding the congested single section travel time to the (negative) exit time, the correct itinerary travel time would be obtained.

An additional remark is that the proposed algorithm cannot split the travel time for the last section "$t_{m-1,m}$" into the main highway trunk travel time and the exit. Nevertheless, this lack of information is not so important, because the last toll plaza is usually located in the main highway trunk and all the vehicles traveling along the last section must go through this toll plaza. In such a way, the interesting information for the driver in this last stretch of the highway is the total aggregated travel time, including both the main trunk travel time and the exit time. In contrast, the exit time in the last toll plaza would be useful information for the highway operator to know the level of service of this last toll plaza.

From the above equations, it can be seen that there are "m-$(i + 1)$" equivalent estimations for the "$i, i + 1$" single section travel time. Each of these estimations results from different itinerary travel times, with its different associated lengths. Note that considering long trips as a possible alternative in the calculation of the single section travel times in Eq. (5.5), implies an increase in the information delay (because the application of Eq. (5.5) requires that all the considered vehicles had left the highway). In addition, travel time for long trips can be considered as less reliable as it can be increased by factors that are unrelated to traffic conditions, for example if some drivers stop for a break or re-fueling. This fact implies an increase in the standard deviation of the average itinerary travel time, due to the higher probability of stops on a long trip. These considerations suggest that for some applications (e.g. real time application) the basic algorithm should be restricted to short trip data. This restriction is detailed in Sect. 4.1, corresponding to real time implementation.

Once all the valid alternatives for estimating the single section travel time are selected (this is different if working off-line with a complete database of events versus working in real time), some smoothing or averaging algorithm must be applied to calculate a unique value for the single section travel time. This smoothing algorithm is presented in Appendix 5C.

5.3.2 Extended Algorithm

Equations (5.5) and (5.6) show that a particular single section travel time is calculated from travel time observations of all the vehicles entering in the origin of the section, except those traveling only in the considered stretch, which are considered in calculating the exit time of the section. The basic algorithm does not consider the vehicles traveling along the stretch but that have entered at a previous entrance.

On certain stretches of the highway, particularly during night hours, the amount of available data may be insufficient to perform the described calculations because of the low flow in a particular itinerary of those considered. Although travel time information under low traffic conditions arouses lesser interest to drivers and highway administrations due to its easier predictability with historic information, in such cases it is possible to increase the amount of available data by considering alternative itineraries for the calculations. For example, a second order algorithm implies the estimation of a two-section (i.e. two consecutive single sections) travel time. If the second order algorithm between entrance 1 and exit 3 is considered, then:

$$t_{1,3} = t_{en(1)} + t_{s(1,3)} + t_{ex(3)} \qquad (5.7)$$

Proceeding in the same way as in Eqs. (5.5) and (5.6), the section travel times and the exit times can be obtained for the (1,3) itinerary as (see Fig. 5.4):

$$t_{s(1,3)} \approx (t_{1,4} - t_{en(1)}) - (t_{3,4} - t_{en(3)}) \approx (t_{1,5} - t_{en(1)}) - (t_{3,5} - t_{en(3)}) \approx \cdots \qquad (5.8)$$

$$t_{ex(3)} = t_{1,3} - t_{s(1,3)} - t_{en(1)} \qquad (5.9)$$

The general expression for a 2nd order algorithm can be written as:

$$\begin{aligned} t_{s(i,i+2)} &\approx (t_{i,i+3} - t_{en(i)}) - (t_{i+2,i+3} - t_{en(i+2)}) \\ &\approx (t_{i,i+4} - t_{en(i)}) - (t_{i+2,i+4} - t_{en(i+2)}) \approx \cdots \end{aligned} \qquad (5.10)$$

$$t_{ex2(i+2)} = t_{i,i+2} - t_{s(i,i+2)} - t_{en(i)} \qquad (5.11)$$

where the subscript "2" in the exit time notation (Eq. 5.11) and in the single section travel time notation (Eq. 5.12), stands for the estimation using a 2nd order algorithm.

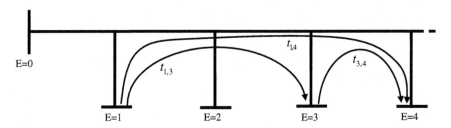

Fig. 5.4 Section (1,3) travel time estimation with second order algorithm

5.3 Estimation of Single Section Travel Times ...

The two-section travel time can be seen as the addition of two consecutive single section travel times

$$t_{s(i,i+2)} = t_{s2(i,i+1)} + t_{s2(i+1,i+2)} \tag{5.12}$$

To calculate these two addends:

$$t_{s2(i,i+1)} = t_{s(i,i+2)} - t_{s(i+1,i+2)} \tag{5.13}$$

where it is assumed that there is enough data to obtain an accurate estimation of the first order single section travel time corresponding to "$t_{s(i+1,i+2)}$". Finally, replacing the result of Eq. (5.13) in Eq. (5.12):

$$t_{s2(i+1,i+2)} = t_{s(i,i+2)} - t_{s2(i,i+1)} \tag{5.14}$$

Equations (5.13) and (5.14) could be easily modified in the case that "$t_{s(i,i+1)}$" was the known addend of Eq. (5.12). If both, "$t_{s(i,i+1)}$" and "$t_{s(i+1,i+2)}$" are known, any combination of Eqs. (5.13) and (5.14) or its modifications could be used (see Appendix 5C for details).

There is one remaining situation, when neither "$t_{s(i,i+1)}$" nor "$t_{s(i+1,i+2)}$" can be obtained from a 1st order algorithm. Then we simply assume the same average speed across both sections, and reach:

$$t_{s2(i,i+1)} = \frac{t_{s(i,i+2)} \cdot l_{s(i,i+1)}}{l_{s(i,i+2)}} \tag{5.15}$$

$$t_{s2(i+1,i+2)} = \frac{t_{s(i,i+2)} \cdot l_{s(i+1,i+2)}}{l_{s(i,i+2)}} \tag{5.16}$$

where "$l_{s(i,j)}$" is the length of the highway stretch between junctions "i" and "j".

This strong assumption implies that the use of Eqs. (5.15) and (5.16) should be restricted to very particular situations where no other information is available.

There still remain some questions to be answered, for example when a first order single section travel time is considered accurate enough, or how to fuse single section travel times (one coming from a first order algorithm and the other from a second order). These aspects are further analyzed in Appendix 5C.

As a summary, the step by step implementation of the running phase of the algorithm (all the default values are already set) results as follows:

- Step 1: Compute all the itinerary travel times in the time interval "p", "$t_{i,j}^{(p)}$" using the filtering algorithm presented in Appendix 5A.
- Step 2: Estimate all the first order single section travel times using all the possible alternatives in Eq. (5.5) and fuse them using the equations presented in

Appendix 5C to find a first order estimation for the single section travel times of all single sections, "$t^{(p)}_{s(i,i+1)}$".

- Step 3: Compute the first order estimation of the exit time "$t^{(p)}_{ex(i+1)}$", using Eq. (5.6).
- Step 4: Estimate all the second order single section travel times using all the possible alternatives in Eq. (5.10) and fuse them using the equations presented in Appendix 5C, to find a second order estimation for the single section travel times of all single sections, "$t^{(p)}_{s2(i,i+1)}$".
- Step 5: Compute the second order estimation of the exit time "$t^{(p)}_{ex2(i+1)}$", using Eq. (5.11).
- Step 6: Fuse the first and second order estimations of the single section travel times and of the exit times using the equations provided in Appendix 5C, to obtain the final travel time estimations.

5.4 Modifications for the Real Time and Off-Line Implementations of the Algorithm

As stated in the first paragraph of Sect. 5.3, any single section travel time "$t_{s(i,i+1)}$" is related to a particular time period "p" (with a duration of "Δt"). Obviously, for the algorithm to be accurate, the average travel times (in particular the ones used in Eqs. (5.5), (5.6), (5.10) and (5.11) must result from observations of vehicles traveling along the same stretch of the highway in the same time period. This has some important implications on the time periods to consider in the different alternatives for estimating a single section travel time (of different itinerary lengths), and are different if working in a real time basis or in an off-line configuration where some future information is available.

5.4.1 Real Time Implementation

For a real time implementation, Eq. (5.5) should be rewritten as:

$$\begin{aligned} t_{i,i+1} = t_{en(i)} + t_{s(i,i+1)} + t_{ex(i+1)} \quad \forall i = 1, \ldots, m-1 \\ t_{i,i+1} = t_{en(i)} + t_{s(i,i+1)} + t_{ex(i+1)} \quad \forall i = 1, \ldots, m-1 \end{aligned} \quad (5.17)$$

In a real time application it is needed to obtain "$i + n*$", the critical highway exit "$i + n$" ($n \geq 2$) which represents the maximum length of an itinerary whose vehicles leaving the highway at time interval "p", and having traveled along the itinerary "$i, i + n*$", have traveled simultaneously with other vehicles traveling from "i" to

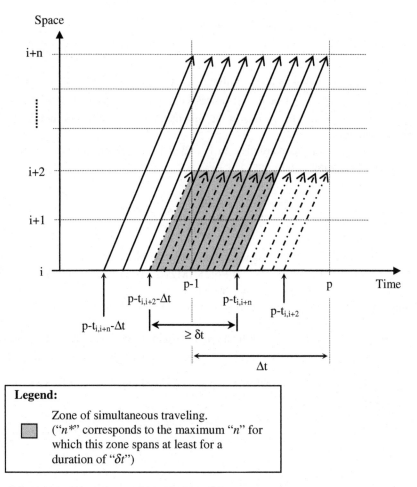

Fig. 5.5 Vehicle trajectories considered in Eq. (5.17)

"$i + 2$" and leaving the highway at the same time interval "p". The simultaneous travel happens along the sections "$i, i + 1$" and "$i + 1, i + 2$". The trajectories diagram sketched in Fig. 5.5 may help to understand this concept.

In Fig. 5.5, "t" represents the required minimum time interval overlapping of trajectories to ensure simultaneous traveling, which is set at 1/3 of "Δt". Note that some additional trajectories in Fig. 5.5 could reach exit "$i + n + 1$" before "p" and still coincide with trajectories traveling along the itinerary ($i, i + 2$). However, this coincide zone will elapse for a time period shorter than the minimum required, "t". Then, the conditions to ensure enough simultaneous traveling can be derived as:

$$\begin{aligned}p - t^{(p)}_{i,i+n} \geq p - t^{(p)}_{i,i+2} - \Delta t + \delta t &\rightarrow t^{(p)}_{i,i+n} \leq t^{(p)}_{i,i+2} + \Delta t - \delta t \\ p - t^{(p)}_{i,i+n} - \Delta t + \delta t \leq p - t^{(p)}_{i,i+2} &\rightarrow t^{(p)}_{i,i+n} \geq t^{(p)}_{i,i+2} - \Delta t + \delta t\end{aligned} \quad (5.18)$$

"$n*$" must be calculated as the maximum "n" for which Eq. (5.18) holds. Note that in general $t^{(p)}_{i,i+2} \leq t^{(p)}_{i,i+n}$ (for $n \geq 2$; unless huge problems take place at exit ramp "$i + 2$") and $\Delta t > \delta t$. This results in the first inequality of Eq. (5.18) being usually the restrictive one.

Applying a similar construction to Eq. (5.10) (2nd order equation), we would obtain the following conditions:

$$\begin{aligned}t^{(p)}_{i,i+n} &\leq t^{(p)}_{i,i+3} + \Delta t - \delta t \\ t^{(p)}_{i,i+n} &\geq t^{(p)}_{i,i+3} - \Delta t + \delta t\end{aligned} \quad (n \geq 3) \quad (5.19)$$

As in general $t^{(p)}_{i,i+3} \geq t^{(p)}_{i,i+2}$ (again unless huge problems take place at exit ramp "$i + 2$"), Eq. (5.19) holds for "$n*$" if Eq. (5.18) does. However, Eq. (5.18) is defined for "$n \geq 3$", which means that the 2nd order travel times can only be considered simultaneously with 1st order travel times in a real time implementation if "$n* \geq 3$".

Perhaps an excess of simplification, but helpful in visualizing the concept, "$t^{(p)}_{i,i+n}$" could be seen as "$t^{(p)}_{i,i+2} + t^{(p)}_{i+2,i+n}$" (the excess of simplification results from the rejection of "$t^{(p)}_{ex(i+2)}$" and "$t_{en(i+2)}$"). Substituting this simplification into Eq. (5.18):

$$\begin{aligned}t^{(p)}_{i+2,i+n} &\leq \Delta t - \delta t \\ t^{(p)}_{i+2,i+n} &\geq -\Delta t + \delta t\end{aligned} \quad (5.20)$$

Equation 5.20 represents the simplified condition to obtain "$n*$", the largest value that can take "n" while not violating the restrictions. Note that the second inequality of this simplified condition is irrelevant as it always holds. "$n*$" sets the possible alternatives in calculating the single section travel times from Eq. (5.17), on a real time basis. These restrictions also have a direct implication in setting "Δt", the larger "Δt", the large "$n*$" could be. This implies more alternatives in estimating the single section travel time, in addition to more observations within each itinerary. In contrast, large "Δt" implies a low updating frequency, which results in an increase of information delay, with disastrous consequences when facing rapidly evolving traffic conditions.

Another implication of these restrictions is the fact that "$n*$" decreases when the travel time in the highway increases (i.e. in congested situations), in particular in highway sections from $(i + 2, i + 3)$ and downstream. If the congested section is the section $(i, i + 1)$ or $(i + 1, i + 2)$, there is no implication related to "$n*$", but this results in an increase of the delay of the information (remember that toll tickets represent a MTT measure).

Finally, it must be considered that the exit time "$t_{ex(i+1)}$" expressed in Eq. (5.6), is also related to a particular time interval "p". Then Eq. (5.6) should be rewritten as:

$$t^{(p)}_{ex(i+1)} = t^{(p)}_{i,i+1} - t^{(p)}_{s(i,i+1)} - t_{en(i)} \qquad (5.21)$$

For Eq. (5.21) to be consistent, the vehicles considered in the calculation of "$t^{(p)}_{s(i,i+1)}$" must have traveled along the section "$i, i+1$" together with those whose average travel time is "$t^{(p)}_{i,i+1}$". Considering the trajectories involved in these calculations, the resulting condition is:

$$\begin{aligned} p - t^{(p)}_{i,i+2} \geq p - t^{(p)}_{i,i+1} - \Delta t + \delta t &\rightarrow t^{(p)}_{i,i+2} \leq t^{(p)}_{i,i+1} + \Delta t - \delta t \\ p - t^{(p)}_{i,i+2} - \Delta t + \delta t \leq p - t^{(p)}_{i,i+1} &\rightarrow t^{(p)}_{i,i+2} \geq t^{(p)}_{i,i+1} - \Delta t + \delta t \end{aligned} \qquad (5.22)$$

Again, in an excess of simplification, "$t^{(p)}_{i,i+2}$" could be seen as "$t^{(p)}_{i,i+1} + t^{(p)}_{i+1,i+2}$". This can be considered approximately true if there's no problem (i.e. congestion) in the entrance or exit ramps at junction "$i + 1$". Substituting this simplification into the first inequality of Eq. (5.22):

$$t^{(p)}_{i+1,i+2} \leq \Delta t - \delta t \qquad (5.23)$$

From this restriction it results that, on a real time information basis, the exit time at ramp "$i + 1$" can only be obtained if Eq. (5.23) holds. Otherwise, the exit time can only be obtained for a past time interval, using the off-line formulation that follows.

5.4.2 Off-Line Implementation

In case the objective is not real time information, and the aim is to reconstruct the single section travel times of past situations (e.g. to obtain travel time templates), then the whole database is available. This means that "future" information (in relation to the time interval of calculation "p") can be used (i.e. there's no need to "wait" until the vehicle has left the highway to obtain its MTT). In this situation, real time restrictions could be modified to obtain a more robust algorithm. Equation 5.17 could be rewritten as (see Fig. 5.6):

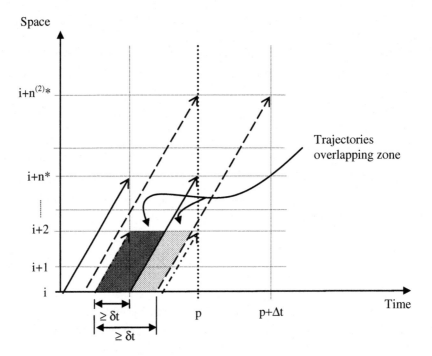

Fig. 5.6 Vehicle trajectories considered in Eq. (5.24)

$$\begin{aligned} t^{(p)}_{s(i,i+1)} &\approx (t^{(p)}_{i,i+2} - t_{en(i)}) - (t^{(p)}_{i+1,i+2} - t_{en(i+1)}) \approx \cdots \\ &\approx (t^{(p)}_{i,i+n*} - t_{en(i)}) - (t^{(p)}_{i+1,i+n*} - t_{en(i+1)}) \\ &\approx (t^{(p+\Delta t)}_{i,i+n*+1} - t_{en(i)}) - (t^{(p+\Delta t)}_{i+1,i+n*+1} - t_{en(i+1)}) \approx \cdots \\ &\approx (t^{(p+\Delta t)}_{i,i+n^{(2)}*} - t_{en(i)}) - (t^{(p+\Delta t)}_{i+1,i+n^{(2)}*} - t_{en(i+1)}) \approx \cdots \end{aligned} \quad (5.24)$$

Using a variation of Eq. (5.18), "$n^{(2)}*$" can be obtained as the maximum "$n^{(2)}$" that fulfills the following restriction:

$$\begin{aligned} t^{(p+\Delta t)}_{i,i+n^{(2)}} &\leq t^{(p)}_{i,i+2} + 2 \cdot \Delta t - \delta t \\ t^{(p+\Delta t)}_{i,i+n^{(2)}} &\geq t^{(p)}_{i,i+2} - 2 \cdot \Delta t + \delta t \end{aligned} \quad (5.25)$$

Equation (5.19) could be modified in the same way to obtain the condition to apply to the 2nd order algorithm.

Then, in a general expression, the single section travel time and the exit time can be calculated off-line as:

5.4 Modifications for the Real Time and ...

$$t^{(p)}_{s(i,i+1)} \approx (t^{(q)}_{i,i+k} - t_{en(i)}) - (t^{(q)}_{i+1,i+k} - t_{en(i+1)}) \quad k = 2, \ldots, m \qquad (5.26)$$

where,

$$q = \begin{cases} p & \text{if } k \leq n* \\ p + \Delta t & \text{if } n* < k \leq n^{(2)}* \\ \vdots & \vdots \\ p + (r-1) \cdot \Delta t & \text{if } n^{(r-1)}* < k \leq n^{(r)}* \end{cases} \quad r \in \mathbb{N} \qquad (5.27)$$

For each "k" it is necessary to find the minimum "$r \in \mathbb{N}$" which satisfies Eq. (5.27), where "$n^{(r)}*$" for a given "r", is the maximum "$n^{(r)}$" that satisfies the following restrictions:

$$\begin{aligned} t^{(p+(r-1)\cdot\Delta t)}_{i,i+n^{(r)}} &\leq t^{(p)}_{i,i+2} + (r-1) \cdot \Delta t - \delta t \\ t^{(p+(r-1)\cdot\Delta t)}_{i,i+n^{(r)}} &\geq t^{(p)}_{i,i+2} - (r-1) \cdot \Delta t + \delta t \end{aligned} \qquad (5.28)$$

Finally, for the off-line calculation of the exit time "$t^{(p)}_{ex(i+1)}$", Eq. (5.21) should be rewritten as:

$$t^{(p)}_{ex(i+1)} = t^{(q)}_{i,i+1} - t^{(q)}_{s(i,i+1)} - t_{en(i)} \qquad (5.29)$$

where "q" is obtained from Eq. (5.27) with a value of "$k = 2$".

5.5 Application to the AP-7 Highway in Spain

This new approach for direct travel time measurement using existing toll infrastructure has been tested for the AP-7 toll highway in Spain. The AP-7 highway runs along the Mediterranean cost corridor, from the French border to the Gibraltar Strait. Nevertheless, the pilot test was restricted to the north east stretch of the highway from "La Roca del Vallès" toll plaza, near Barcelona, to the French border at "La Jonquera". This stretch is approximately 120 km long.

The first of the pilot tests was performed with the April 18th 2008 data. This was a very conflictive Friday in terms of traffic, as a fatal crash happened on the highway. The accident forced the closing of two of the three existing lanes in the southerly direction towards Barcelona, causing severe congestion and serious delays.

Figure 5.7 shows the physical configuration of the test site and the spatial and temporal evolution of the traffic congestion on the highway (shock-wave analysis). From the diagram, it can be seen that the accident happened around 3.00 pm near the "Cardedeu" junction, causing a huge bottleneck at this location. The accident could not be cleared until 5.00 pm, causing long queues to grow (longer than 13 km).

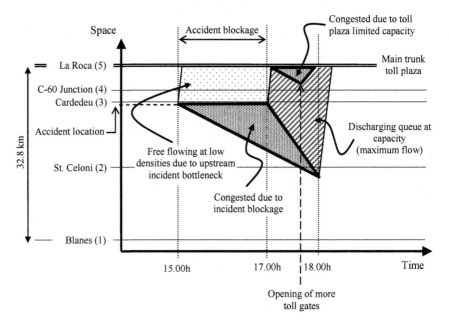

Fig. 5.7 Space-time evolution of traffic states in the highway (April 18th 2008)

After the accident was cleared (and the corresponding bottleneck removed), the queue discharged at capacity, with a flow in excess of the maximum service rate of the available toll gates at the "La Roca" toll plaza, located on the main highway trunk. This caused a small queue to grow just upstream of the toll barrier, which started to dissipate when the capacity of the toll plaza was increased by the opening of more toll gates. The whole incident implied a maximum delay of 1 h for the vehicles travelling from "Blanes" to "La Roca" (a distance of 32.8 km), as can be seen in Fig. 5.8. This conflictive situation with rapidly evolving conditions represents a perfect environment for testing the proposed algorithm, as travel time information is crucial and the delay in reporting the information is disastrous. In addition, in order to fully test the algorithm for different conditions on the highway, in particular for more "normal" conditions, a second pilot test is presented with April 27th 2008 data, a usual Sunday in the highway with recurrent evening slight delays on the southbound direction (see Fig. 5.8), resulting from the massive return to Barcelona after spending a day or weekend on the coast.

The present application of the algorithm will be performed in an on-line basis. This selection is due to the fact that the on-line application is more restrictive than the off-line, and the contribution of the method is more relevant, because in addition to provide detailed decomposition of itinerary travel time in single sections and exit times (like in the off-line application), it also provides an increased immediacy in reporting travel time information to the drivers, crucial in real-time information systems.

5.5 Application to the AP-7 Highway in Spain

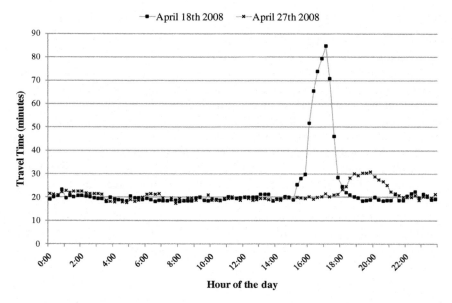

Fig. 5.8 Travel time from Blanes to La Roca. *Note* These travel times correspond to "$t_{1,5}^{(p)}$", the representative average of the MTT for the itinerary "1,5" every 15 min directly obtained from toll ticket data

5.5.1 Selection of the Sampling Duration "Δt"

The selection of the sampling duration "Δt" is, in practice, very much relevant, as it determines how well the method performs with data typically available. Two goals must be pursued in the selection of "Δt". On the one hand "Δt" should be as short as possible in order to provide a frequent update and an accurate tracking of travel time evolution, avoiding averaging and smoothing the possible rapidly changing traffic conditions. On the other hand, "Δt" should be large enough to include enough travel time observations for the estimation to be statistically significant. From basic estimation theory it can be stated that, given a desired statistical significance or probability level of the estimation, "$\alpha \in (0, 1)$", the resulting maximum absolute error "ε" in the estimation of an average travel time "$t_{i,j}^{(p)}$" from individual measurements, assumed independent, is related to the number of observations "$N_{i,j}^{(p)}$" by:

$$\varepsilon_{(t_{i,j}^{(p)})} = \frac{Z_{(1+\alpha)/2} \cdot \sigma_{(t_{i,j}^{(p)})}}{\sqrt{N_{i,j}^{(p)}}} \qquad (5.30)$$

where "$\sigma_{(t_{i,j}^{(p)})}$" stands for the standard deviation of travel time observations and where applying the standard normal cumulative distribution function to "$Z_{(1+\alpha)/2}$" we obtain the cumulative probability of "$(1+\alpha)/2$".

A minimum "N" must be achieved in the itinerary travel time estimation in order to delimit the maximum error, given a statistical significance. In general, for a given demand on the highway, the maximum error will be lower for longer "Δt"s, as more observations will be available. In addition, for a given "Δt", the error will be lower in those situations with high demand and low travel time variance.

The error formulated in Eq. (5.30), is not directly the absolute error in the estimation of single section travel times from the proposed algorithm. Note that, in the on-line application of the algorithm, the single section travel time results from the average of "$n* - 1$" subtractions between two different itinerary travel times (see Eq. 5.17). In addition, if "$n* \geq 3$", "$n* - 2$" additional estimations are available from a 2nd order algorithm (see Eq. 5.19). Then it can be stated that the maximum absolute error in the estimation of a single section travel time, "$t_{s(i,j)}^{(p)}$", is:

$$\varepsilon_{(t_{s(i,i+1)}^{(p)})} = \frac{\sqrt{2} \cdot \varepsilon_{(t_{i,j}^{(p)})}}{\sqrt{n*-1}} \quad \text{if } n* = 2$$

$$\varepsilon_{(t_{s(i,i+1)}^{(p)})} = \frac{\sqrt{2} \cdot \varepsilon_{(t_{i,j}^{(p)})}}{\sqrt{2n*-3}} \quad \text{if } n* \geq 3$$

(5.31)

Longer "Δt"s imply an increase in the number of the itineraries considered in the calculation of "$t_{s(i,j)}^{(p)}$" in the case of an on-line application (i.e. an increase in "$n*$", see Eq. 5.18), and therefore a second source for the error reduction when increasing "Δt". For the off-line application the number of considered itineraries is maximum and independent of "Δt" (i.e. "$n* = m$"). Then, in general, for a same "Δt" the obtained maximum error of the single section travel time off-line estimation will be lower or equal than in the on-line case.

In the selection process for "Δt", four options are considered: 1, 3, 5 and 15 min. 1 min is considered to be a lower bound as it does not seem necessary to increase further the updating frequency because traffic conditions do not evolve so quickly. In addition, and as it will be seen next, the proposed algorithm may not be suitable to work with this fine resolution. 15 min is considered as an upper bound because it is considered to be the minimum updating frequency acceptable in order to be used as a real time information system. Table 5.1 shows the expected maximum absolute error obtained for different "Δt"s and considering the typical demand patterns in the test site.

If travel time information is disseminated every two single sections (e.g. using variable message signs), from Table 5.1 it can be seen that the expected average cumulative error in two sections for an updating interval of 5 min is below the one minute error threshold for at least half of the calculation periods in almost all demand patterns. Although the selection of a 5 min "Δt" could be used for all

5.5 Application to the AP-7 Highway in Spain

Table 5.1 Maximum expected estimation errors for various demand patterns and different "Δt"

Demand pattern[a]	Δt (min)	Itinerary travel time standard deviation "σ" (min)		N		n^*		Maximum expected single section travel time absolute error[b] (min)	
		Median	25–5 % Percentile	Median	25–75 % Percentile	Median	25–75 % Percentile	Median	25–75 % Percentile
A	1	1.11	0.63–1.70	2	1–3	2	2–2	1.22	0.52–2.40
	3	1.29	0.92–1.85	3	1–6	2	2–2	1.00	0.51–2.62
	5	1.37	0.98–1.87	4	1–9	4	3–4	0.45	0.21–1.33
	15	1.55	1.55–1.87	22	11–33	4	4–4	0.21	0.14–0.36
B	1	1.22	0.92–1.56	4	1–9	2	2–2	0.90	0.44–2.20
	3	1.24	0.90–1.47	5	2–22	2	2–2	0.81	0.27–1.61
	5	1.22	0.87–1.49	7	2–30	4	3–4	0.30	0.10–0.94
	15	1.39	1.30–1.54	114	58–151	4	4–4	0.08	0.07–0.13
C	1	0.99	0.73–1.26	7	2–12	2	2–2	0.53	0.29–1.38
	3	0.99	0.73–1.24	8	2–34	2	2–2	0.49	0.18–1.36
	5	1.00	0.69–1.26	9	2–52	4	3–4	0.21	0.06–0.72
	15	1.18	1.02–1.28	181	114–218	4	4–4	0.06	0.04–0.08
D	1	3.09	2.09–3.95	6	1–10	2	2–2	1.84	0.92–4.83
	3	2.76	1.15–3.65	7	2–28	2	2–2	1.47	0.31–3.65
	5	2.46	0.93–3.61	9	2–42	3	3–3	0.68	0.12–1.93
	15	3.57	3.09–4.06	134	110–167	4	4–4	0.2	0.15–0.25

(continued)

Table 5.1 (continued)

Demand pattern[a]	Δt (min)	Itinerary travel time standard deviation "σ" (min)		N		n^*		Maximum expected single section travel time absolute error[b] (min)	
		Median	25–5 % Percentile	Median	25–75 % Percentile	Median	25–75 % Percentile	Median	25–75 % Percentile
E	1	2.19	1.48–3.38	4	1–7	2	2–2	1.55	0.79–4.78
	3	2.02	1.24–2.93	5	2–18	2	2–2	1.28	0.41–3.21
	5	1.98	1.24–2.85	7	2–28	4	3–4	0.48	0.15–1.64
	15	2.60	2.14–4.21	89	74–104	4	4–4	0.17	0.13–0.31

Note
[a]Data obtained from April 18th and 27th, 2008
A. Night hours (low demand). Average itinerary travel time considering all itineraries between junctions (1) and (5) = 6.54 min
B. Off-Peak hours (moderate demand). Average itinerary travel time = 6.40 min
C. Peak hours (free flowing high demand). Average itinerary travel time = 5.91 min
D. Recurrent congested periods (high demand). Average itinerary travel time = 11.17 min
E. Incident conditions (moderate demand + congestion). Average itinerary travel time = 12.37 min
[b]Applying Eqs. (5.30) and (5.31), considering a probability level of $\alpha = 0.68$ and the average values for "N" and "σ" shown in this table

5.5.2 Accuracy of the Algorithm

To check the accuracy of the algorithm, the ground truth travel times experienced by the drivers travelling from "Blanes" to "La Roca", "$t_{1,5}^{(p)}$" obtained as an accurate average of the itinerary travel times resulting from the toll ticket data of only those vehicles travelling in that particular itinerary and reaching (5) in any "p" time interval (this data is plotted in Fig. 5.8, considering "$\Delta t = 15$ min"), will be compared with "$\tilde{t}_{(1,5)}^{(p)}$" the travel time resulting from the proposed algorithm (i.e. the addition of single section travel times, entrance time and exit time) along the same itinerary and exiting at the same time interval "p".

In order to evaluate the performance of the algorithm in terms of accuracy, the travel times to compare must be of the same nature. Then, as "$t_{1,5}^{(p)}$" is an MTT, "$\tilde{t}_{(1,5)}^{(p)}$" must be so. This means that "$\tilde{t}_{(1,5)}^{(p)}$" does not result from the simple addition of single section travel times at time interval "p", but from a backwards trajectory reconstruction process. This is:

$$\tilde{t}_{(1,5)}^{(p)} = t_{en(1)} + \sum_{i=1}^{4} t_{s(i,i+1)}^{(q_i)} + t_{ex(5)}^{(p)} \qquad (5.32)$$

where "q_i" stands for the time period to consider in the trajectory reconstruction, as it will vary taking into account that the time taken up by the virtual vehicle in travelling along consecutive sections must be considered when trying to estimate a trajectory based travel time (e.g. MTT or PTT). In contrast, in case of estimating an ITT, which is not trajectory based, these "q_i"s time periods have to be considered all equal to "p". In the present case, a backwards reconstruction is needed and "q_i"s have to be calculated iteratively starting from downstream.

Although the single section travel times that configure the itinerary will only vary every "Δt", the headway between launched virtual vehicles can be as short as desired. In the present application, one virtual vehicle was launched every minute.

Table 5.2 shows the error committed in the estimation of "$\tilde{t}_{(1,5)}^{(p)}$" in relation to the ground truth "$t_{1,5}^{(p)}$" for several demand patterns from different days, and considering various "Δt". Note from the numerical values that, in general, the obtained error follows the logic detailed in the previous section for the expected single section estimation error. However it is worth to notice that for the rapidly evolving incident related traffic conditions (i.e. scenario E in Table 5.2) the obtained errors for "$\Delta t = 15$" are significantly larger than expected. This is a clear consequence of the

Table 5.2 Accuracy of the algorithm for various demand patterns and different "Δt" considering the itinerary (1,5)

Demand pattern[a]	Test day	Period of the day	Δt (min)	Itinerary travel time absolute error[b] (relative)	
				Mean	Max.
A	April 18th, 2008	0 am–6 am 9 pm–12 pm	3	1.44 (0.07)	6.25 (0.25)
			5	1.18 (0.06)	3.14 (0.17)
			15	1.03 (0.05)	2.48 (0.13)
	April 27th, 2008	0 am–8 am	3	1.74 (0.09)	4.19 (0.24)
			5	1.66 (0.08)	4.49 (0.19)
			15	1.62 (0.08)	3.61 (0.16)
B	April 18th, 2008	6 am–7 am 9 am–3 pm 7 pm–9 pm	3	0.95 (0.05)	4.59 (0.20)
			5	1.07 (0.06)	3.53 (0.21)
			15	0.94 (0.05)	2.30 (0.12)
	April 27th, 2008	8 am–4 pm 10 pm–12 pm	3	1.09 (0.06)	3.43 (0.17)
			5	1.14 (0.06)	3.26 (0.17)
			15	0.89 (0.05)	3.34 (0.17)
C	April 18th, 2008	7 am–9 am	3	1.30 (0.07)	3.32 (0.18)
			5	1.04 (0.06)	2.87 (0.13)
			15	0.75 (0.04)	1.08 (0.06)
	April 27th, 2008	4 pm–5 pm	3	0.87 (0.04)	2.89 (0.14)
			5	0.65 (0.03)	1.56 (0.08)
			15	0.76 (0.04)	1.08 (0.05)
D	April 27th, 2008	5 pm–10 pm	3	1.47 (0.06)	5.07 (0.16)
			5	1.28 (0.05)	4.25 (0.14)
			15	1.06 (0.04)	2.30 (0.09)
E	April 18th, 2008	3 pm–7 pm	3	3.62 (0.09)	14.72 (0.34)
			5	3.59 (0.08)	11.58 (0.23)
			15	5.84 (0.11)	18.09 (0.30)

Note [a]Periods defined as in Table 5.1
[b]In minutes

lack of independence between travel time individual measurements in a 15 min period during rapid congestion onset or dissolve.

From these results it can be confirmed that for free-flowing conditions (i.e. demand scenarios A-C) and even for moderate recurrent congestion (i.e. scenario D) an updating interval of 15 min provides the best accuracy results. However, this long interval is not capable of tracking the rapidly changing conditions of incident related congestion (i.e. scenario E), particularly in the congestion onset. For this last situations, "$\Delta t = 5$ min" is adequate. Figure 5.9 provides graphical evidence of the algorithm behavior in congested conditions.

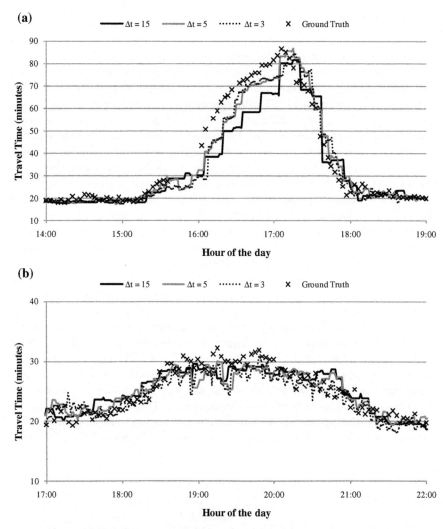

Fig. 5.9 Accuracy of the algorithm for different updating periods. **a** Incident related congestion (April **18th**, 2008). **b** Moderate recurrent congestion (April 27th, 2008)

An updating interval of 5 min can be selected as a compromise solution for all the demand scenarios. In this case, the algorithm is accurate enough to provide travel time information to drivers and road administrations, with a mean absolute relative error below 10 % in all situations, and in particular, in critical situations with huge and rapid variations in travel times.

5.5.3 Value as a Real Time Information System

One of the main advantages of the estimation of single section travel times is the reduction of the information delay for long itineraries when providing real time information to drivers. This improvement, resulting from the dissemination of an ITT for the itinerary (instead of a MTT with a greater delay), is only relevant when traffic conditions evolve in a time horizon equal to the travel time (i.e. congestion onset and congestion dissolve). Otherwise, both estimations would lead to the same result.

Table 5.3 shows the travel times to be disseminated in real time for the same analyzed itinerary (1, 5) in the case of using the information directly obtained from toll ticket data (MTT) or the ITT obtained by the addition of single section travel times resulting from the proposed algorithm. Both are compared with the real travel time of drivers *entering* the highway during the next time period, who are able to receive the information (note that this is the target information—a PTT—which is not known at the instant the information is disseminated. This target travel time is only known after the vehicles have left the highway).

As can be seen from results in Table 5.3 from data corresponding to both test days and different evolving traffic conditions, ITT performs well in recurrent and non-recurrent traffic conditions. Generally performs better than MTT, particularly under non-recurrent and rapidly changing conditions, where the benefits of reducing the information delay are bigger. However, the same quickly evolving conditions also imply big differences between ITT and PTT. Therefore one could say that while the performance of MTT in supporting real time information systems is bad, the performance of ITT is not as bad, but still flawed due to the null forecasting capabilities of a real time measurement. This is the most that can be done without sinking in the uncertainties of forecasting and modeling. These concepts are clarified in Fig. 5.10, using a trajectories diagram.

Table 5.3 Real time travel time dissemination in relation to the available information

Type of available information	Travel time disseminated at "p" (min:sec)	
	April 18th 2008 "p" = 17.30 h (scenario E) Congestion dissolve	April 27th 2008 "p" = 18.15 h (scenario D) Congestion onset
Only itinerary information (MTT) $t_{1,5}^{(p)}$	46:04	24:32
Single section travel time information (ITT) $t_{en(1)}^{(p)} + \sum_{i=1}^{4} t_{s(i,i+1)}^{(p)} + t_{ex(5)}^{(p)}$	37:18	27:16
Real travel time for those vehicles entering the highway between instants "p" and "$p+1$" (PTT)	24:47	30:14

5.5 Application to the AP-7 Highway in Spain

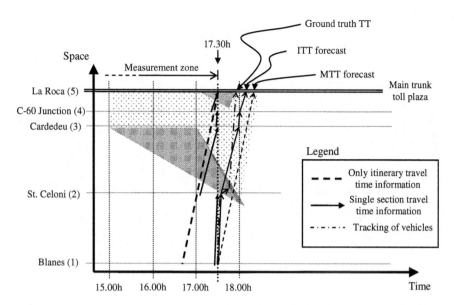

Fig. 5.10 Effect of the available information in the dissemination of travel times (April 18th, 2008)

5.5.4 Exit Time Information

The proposed algorithm for estimating single section travel times in a closed highway system provides, as a collateral result, the highway exit time. This information is of great interest to highway operators, since drivers' value of this particular time is greater than the traveling time, strictly speaking. This means that a little delay in the payment at the toll gate is a big nuisance to drivers, who attribute a low level of service to the highway trip. Obviously these delays for payment greatly penalize the operators' reputation.

For this reason, highway operators are especially sensitive to keep this exit time as low as possible, while maintaining costs. This trade-off is achieved with an accurate schedule of toll gates (i.e. number of open toll gates, and direction of operation, as some of the toll gates are reversible). In fact, the experience achieved during years of operation on the AP-7 highway has given the operators an accurate knowledge of the demand pattern at each junction, so that in most of the situations all the exit times are limited to between 1 and 3 min, which represent very short delays for exiting the highway. However, in some incident situations, with unexpected demand, or some problems at the toll gates, queues could grow, causing delays to be of more consideration (see Fig. 5.11).

Under these conditions, a quantitative estimation of the exit times at the junction can alert the highway operations center in order to reschedule the toll gates. In addition, information could be disseminated to drivers to allow them to select an

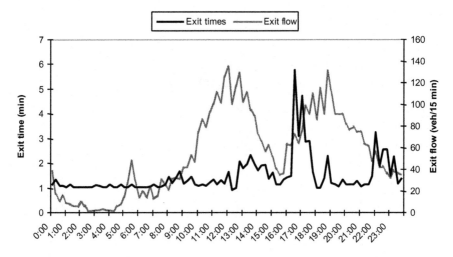

Fig. 5.11 Exit times at the Cardedeu (3) junction, (April 6th 2008)

alternative junction to exit the highway, avoiding a delay in the trip and helping to alleviate the problem.

5.6 Conclusions and Further Research

Link travel time is the most appreciated information for road users. The new approach in this chapter for calculating the travel time on highways using toll infrastructure is a simple one and can be easily put into practice with the existing infrastructure. The scheme uses data obtained from the toll tickets on highways with a closed tolling system. Rather than simply calculating the itinerary travel time by comparing the entry and exit times on the ticket, the approach presented here could be used to increase the information available from toll ticket data.

The proposed method is capable of estimating single section travel times (i.e. time required to travel between two consecutive junctions on the main trunk of the highway) and also the exit time at each junction (i.e. the time required to travel along the exit link plus the time required to pay the fee at the toll gate). Combining both estimations it is possible to calculate all the required itinerary travel times, even those with very few observations where direct measurement would be problematic, and avoiding the information delay for real time application.

The results of the pilot test carried out on the AP-7 highway in Spain indicate the suitability of the method for the link travel time estimation in a closed toll highway system.

Moreover, the accuracy in the travel time estimation should make the development of a robust incident detection system possible, by comparing the real time estimations to the recurrent travel times. Since the method supplies exit time information, it is also possible to detect whether the incident was on the main highway or in the ramp toll plaza. This is valuable information for the highway operators, enabling them to deal with the incident.

Further research may consider data fusion with other sources of data, such as loop detectors, to obtain information within the inner section (between junctions). Travel time prediction (to improve the forecasting capabilities of a real time measurement) on the basis of the present scheme is also a key factor for future research.

Appendix 5A: Obtaining a Representative Average of Itinerary Travel Times in a "Δt" Time Interval

The source data to calculate the single section travel times using the proposed algorithm is the "$t_{i,j}^{(p)}$", symbolizing a representative average of the individual itinerary travel time observations "$t_{i,j,k}^{(p)}$" obtained in the "p" time interval (i.e. all the "k" vehicles that have exited the highway at "j", coming from "i" between the time instants "$p - \Delta t$" and "p"). As introduced in Sect. 5.3, the estimation of this average can be tricky. In fact, it represents the main goal in the ETC-only based travel time estimation systems, where vehicles are identified at several control points on the main highway trunk, and the measured travel times are directly single section travel times.

Problems in the estimation of a representative average of itinerary travel times arise from the following characteristics of these data (see Fig. 5A.1):

- Possibility of high variability in travel time observations within a "Δt" time interval (higher variability as "Δt" is higher), and high variability in travel time observations between consecutive time intervals.
- Few observations for some combinations of itinerary and time interval.
- Presence of outliers, whose measured travel time is not related to traffic conditions. There are two main types of outliers: travel time is in excess due to stops on the highway during the trip, and travel time is overly low resulting from motorbikes dodging traffic jams.

Obviously, outliers must not be considered in the estimation of the average travel time. However, it is not obvious to decide on whether an observation is an outlier or not (unless it is very extreme). This is particularly difficult in the case of very few observations, and taking into account the possibility of high variability in travel times (e.g. an accident has happened).

In this situation, standard methods for outlier identification are not suitable, and a smart data filtering process is necessary, taking into account the traffic state. The data filtering method presented in this appendix uses two alternatives, depending on the number of observations in the set. On the one hand, if the set contains enough observations, the median represents a good average and only a smoothing process is

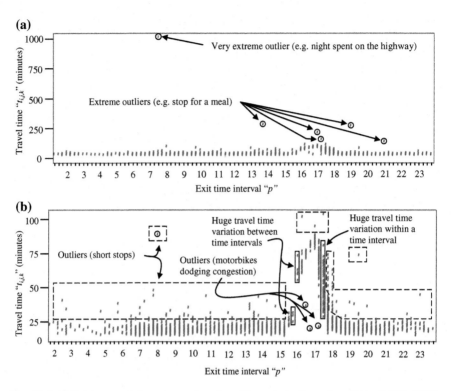

Fig. 5A.1 Toll ticket travel time observations from Blanes to La Roca, "$t_{1,5,k}$" (April 18th 2008). **a** All the observations. **b** Truck observations and extreme outliers removed

required. On the other hand, for sets with few observations, a more careful filtering is needed.

5A.1 Elimination of Extreme Outliers

Extreme outliers or measurement errors are easily detected (see Fig. 5A.1), and therefore can be easily eliminated from the observations database. An observation "$t_{i,j,k}$" is considered as an extreme outlier if:

$$\begin{aligned} & t_{i,j,k} < 0 \\ & t_{i,j,k} > 5 \text{ h} \\ & i = j \\ & \frac{l_{i,j}}{t_{i,j,k}} > 200 \text{ km/h} \end{aligned} \quad (5A.1)$$

where "$l_{i,j}$" is the total length of the itinerary (considering the on and off ramps).

For a maximum length of itinerary of about 120 km, all the situations captured by Eq. (5A.1) represent measurement errors or stops. However, not all the measurement errors or stops are captured by Eq. (5A.1).

5A.2 Data Filtering in Sets with Enough Data

Assume that a set of itinerary travel times, defined by an origin "i", a destination "j" and a time interval "p", has "$N_{i,j}^{(p)}$" observations. This set has enough data if "$N_{i,j}^{(p)}$" is bigger than a threshold value "$N_{i,j}^{*(p)}$", which must be set for the selected "Δt" using Eq. (5.30), data in Table 5.1, and defining a maximum acceptable error given a statistical significance for the estimation. For instance, in the present application for "$\Delta t = 5$ min", considering a maximum acceptable error of 1 min in the average itinerary travel time estimation with a statistical significance "$(1 - \alpha)$" of 0.9, the threshold value "$N_{i,j}^{*(p)}$" equals 8 itinerary travel time observations. In this case where the set has enough data, "$\hat{t}_{i,j}^{(p)}$" is defined as the median of the "$t_{i,j,k}$" observations, where "$k = 1, \ldots, N_{i,j}^{(p)}$". The interquartile range of the set is also computed, "$I\hat{Q}R_{i,j}^{(p)} = \hat{q}_3 - \hat{q}_1$", where "$\hat{q}_3$" and "$\hat{q}_1$" are the 75 and 25 % percentiles respectively. In the present case, with enough observations, it is not necessary to explicitly eliminate the outliers, as the problem is solved by considering the median statistic (instead of the mean), which is highly insensitive to a few extreme values.

The representative average to use in the single section travel time estimation algorithm is an exponentially smoothed value of "$\hat{t}_{i,j}^{(p)}$". The exponential smoothing is carried out on a logarithmic scale (Eq. 5A.2) to account for the log-normal distribution of travel times (Li et al. 2007). Note that travel time distribution is skewed to the right, reflecting the fact that travel times do not significantly decrease below a free flowing travel time (vehicles travelling around the posted speed limit), while significantly higher travel times are possible, especially if congestion builds up. This suggests that the lower "Δt" is, the lower will be this skewness (for small "Δt"s it is less likely that different traffic states arise), and therefore, travel time distribution would tend to normal for small time intervals of measurement.

$$\begin{aligned} t_{i,j}^{(p)} &= \exp\left[\alpha_{i,j}^{(p)} \ln\left(\hat{t}_{i,j}^{(p)}\right) + \left(1 - \alpha_{i,j}^{(p)}\right) \cdot \ln\left(t_{i,j}^{(p-1)}\right)\right] \\ q_{1\,i,j}^{(p)} &= \exp\left[\alpha_{i,j}^{(p)} \ln\left(\hat{q}_{1\,i,j}^{(p)}\right) + \left(1 - \alpha_{i,j}^{(p)}\right) \cdot \ln\left(q_{1\,i,j}^{(p-1)}\right)\right] \\ q_{3\,i,j}^{(p)} &= \exp\left[\alpha_{i,j}^{(p)} \ln\left(\hat{q}_{3\,i,j}^{(p)}\right) + \left(1 - \alpha_{i,j}^{(p)}\right) \cdot \ln\left(q_{3\,i,j}^{(p-1)}\right)\right] \\ IQR_{i,j}^{(p)} &= q_{3\,i,j}^{(p)} - q_{1\,i,j}^{(p)} \end{aligned} \quad (5A.2)$$

The statistical significance of the average itinerary travel time estimation, "$\alpha_{i,j}^{(p)} \in (0, 1)$", is selected as the smoothing factor. "$\alpha_{i,j}^{(p)}$" depends on the number of observations in the set and on the acceptable absolute error "ε", which may be used as a calibration parameter. In the present application "ε" has been set to 1 min, As there are fewer observations, the average will be less reliable, "$\alpha_{i,j}^{(p)}$" will be lower and greater weight will be given to past observations. "$\alpha_{i,j}^{(p)}$" could be seen as a reliability index for "$t_{i,j}^{(p)}$", taking a value of one for perfect information given the acceptable error and decreasing as the information becomes less reliable. A minimum weight of 0.5 is given in the case where there is at least one valid observation. More formally:

$$\alpha_{i,j}^{(p)} = \max \left\{ 2 \cdot F^{-1}\left(\frac{\varepsilon \cdot \sqrt{N_{i,j}^{(p)}}}{\sigma_{(t_{i,j}^{(p)})}} \right) - 1, \ 0.5 \right\} \qquad (5A.3)$$

where "$F^{-1}(z)$" is the inverse cumulative distribution function of a standard normal probability distribution and "$\sigma_{(t_{i,j}^{(p)})}$" is the standard deviation of the average itinerary travel time estimation. Approximate average data for this standard deviation is provided in Table 5.1.

In the present section, dealing with groups with enough observations, "$\alpha_{i,j}^{(p)}$" is defined by the first condition of Eq. (5A.3), and the result is approximately equal to one (perfect information). In this case, Eq. (5A.2) is of little use as "$t_{i,j}^{(p)} \approx \hat{t}_{i,j}^{(p)}$".

5A.3 Data Filtering in Sets with Few Data

In case the set does not have enough data ("$0 < N_{i,j}^{(p)} < N_{i,j}^{*(p)}$"), then the median of these observations cannot be considered a representative average of the itinerary travel time for that time interval. The data filtering process in this case, tries to decide if an observation could be an outlier or not. If the answer is positive, then the observation is eliminated from the database. The median of the remaining valid observations is set as "$\hat{t}_{i,j}^{(p)}$", the input required for the exponential smoothing process detailed in Eqs. (5A.2) and (5A.3).

To determine if an observation could be an outlier, two confidence intervals are defined: $\left(t_{\min i,j}^{(p)}, t_{\max i,j}^{(p)} \right)$ and a broader $\left(t_{MIN i,j}^{(p)}, t_{MAX i,j}^{(p)} \right)$, where:

$$t_{\max i,j}^{(p)} = \exp\left[\ln\left(q_{3\,i,j}^{(p-1)}\right) + \gamma \cdot \ln\left(IQR_{i,j}^{(p-1)}\right)\right]$$

$$t_{MAX i,j}^{(p)} = \exp\left[\ln\left(q_{3\,i,j}^{(p-1)}\right) + \rho_{i,j}^{(p-1)} \cdot \ln\left(IQR_{i,j}^{(p-1)}\right)\right]$$

$$t_{MINi,j}^{(p)} = \exp\left[\ln\left(t_{i,j,0}\right) - \gamma \cdot \ln\left(IQR_{i,j,0}\right)\right] \tag{5A.4}$$

$$t_{\min i,j}^{(p)} = \begin{cases} \exp\left[\ln\left(q_{1\,i,j}^{(p-1)}\right) - \gamma \cdot \ln\left(IQR_{i,j}^{(p-1)}\right)\right] & \text{if } \geq t_{MINi,j}^{(p)} \\ t_{MINi,j}^{(p)} & o.w. \end{cases}$$

These confidence intervals for an itinerary "i, j" and time interval "p" are defined as an extra time above the third quartile (maximum) or below the first quartile (minimum) of travel time distribution in the previous time interval. These extensions depend on the interquartile range of travel times in the previous time interval. This responds to the fact that when some change starts to develop in a "p" time period, the interquartile range increases. Then the confidence intervals for the possible acceptance of an observation for the next time period will be broader, with larger variations than usual in travel times. The lowest threshold of these confidence intervals is limited by "$t_{MINi,j}^{(p)}$", computed as the free flowing travel time "$t_{i,j,0}$" (defined as the first quartile of the travel time distribution in free flowing conditions) minus a fraction of the free flowing conditions interquartile range "$IQR_{i,j,0}$". Obviously, none of these thresholds can exceed the extreme values set in Eq. (5A.1). Only as an order of magnitude, "$IQR_{i,j,0}$" results approximately from a variation interval of 20 km/h around the average free flow speed of approximately 110 km/h. The absolute magnitude of "$IQR_{i,j,0}$" depends on the length of the "i,j" itinerary.

The amplitudes of the confidence intervals defined in Eq. (5A.4), depend on a proportionality constant of "γ", calibrated to 0.5, and on "$\rho_{i,j}^{(p-1)}$" for the "MAX" threshold. "$\rho_{i,j}^{(p-1)}$" is defined as:

$$\rho_{i,j}^{(p-1)} = \lambda \cdot \left[2 - \left(\alpha_{i,j}^{(p-1)}\right)^{\max(N_{out\,i,j}^{(p-1)},\,s_{i,j}^{(p-1)})}\right] \tag{5A.5}$$

where "$\alpha_{i,j}^{(p-1)} \in (0.5, 1)$", "$N_{out\,i,j}^{(p-1)}$" is the number of observations that have been considered outliers in the "$p - 1$" time interval, and "$s_{i,j}^{(p-1)}$" is the number of consecutive intervals without any observation before the time interval "$p - 1$". The larger these last two variables are, the less reliable the previous travel time average is considered to be, as it is not tracking accurately the travel time evolution. As a consequence, broader confidence intervals will be considered for the target time interval "p". "λ" is the default value for perfect previous information and is set to a value of 3.

In this context, four situations can be defined, given a travel time observation "$t_{i,j,k}^{(p)}$":

(a) $t_{i,j,k}^{(p)} \in \left[t_{\min i,j}^{(p)}, t_{\max i,j}^{(p)} \right]$; Then "$t_{i,j,k}^{(p)}$" is considered a valid observation.

(b) $t_{i,j,k}^{(p)} \notin \left[t_{MINi,j}^{(p)}, t_{MAXi,j}^{(p)} \right]$; Then "$t_{i,j,k}^{(p)}$" is considered an outlier and is eliminated from the database.

(c) $t_{i,j,k}^{(p)} \in \left[t_{\max i,j}^{(p)}, t_{MAXi,j}^{(p)} \right]$; Then "$t_{i,j,k}^{(p)}$" is considered a Type 1 doubtful observation.

(d) $t_{i,j,k}^{(p)} \in \left[t_{MINi,j}^{(p)}, t_{\min i,j}^{(p)} \right]$; Then "$t_{i,j,k}^{(p)}$" is considered a Type 2 doubtful observation.

5A.3.1 Deciding on Type 1 Doubtful Observations

An itinerary travel time observation falling in the excess doubtful zone can result from two situations:

- The vehicle has stopped for a short time (e.g. for a quick refuel). Then the observation should be considered as an outlier and eliminated from the database.
- There is a sudden travel time increase in the highway (e.g. due to an incident). Then the observation should be considered as valid.

To decide which is the cause of this doubtful observation, two contrasts are developed. Firstly, the difference between the doubtful observation and the other existing valid observations in the set is analyzed (overtaking contrast). If this difference is considered small, the observation is accepted. Otherwise, if the previous contrast cannot be applied (e.g. there are few valid observations in the set) or the difference is considered to be rather large within a time interval, then the traffic state (i.e. congested or not) is assessed. If traffic is considered to be congested, the observation is considered as valid, otherwise, the observation is an outlier, and it is eliminated from the database.

Overtaking Contrast
This contrast stands for the fact that if some vehicles are capable of achieving significantly lower travel times for the same time interval and itinerary as the doubtful observation, then these vehicles are overtaking the "doubtful" vehicle whose large travel time is not related to general traffic conditions but to the specific behavior of this vehicle. Specifically, assume that at least one valid observation exists in the set of observations where the doubtful "$t_{i,j,k}^{(p)}$" is contained. Name vehicle "l" the vehicle whose itinerary travel time, "$t_{i,j,l}^{(p)}$" is the maximum within the valid observations, then "$t_{i,j,k}^{(p)}$" is also considered valid if:

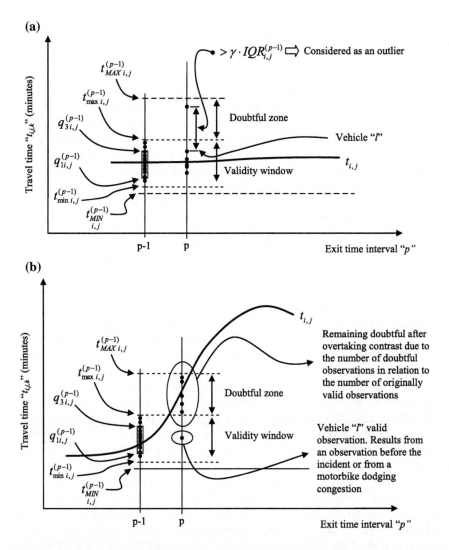

Fig. 5A.2 Sketch of the overtaking contrast. **a** Elimination of an outlier. **b** Overtaking contrast does not make it possible to decide

$$t_{i,j,k}^{(p)} \leq t_{i,j,l}^{(p)} + \gamma \cdot IQR_{i,j}^{(p-1)} \tag{5A.6}$$

If Eq. (5A.6) does not hold and the number of originally valid observations in the set exceeds "$N_{i,j}^{(p)}/2$" (i.e. half of the total number of observations, then "$t_{i,j,k}^{(p)}$" is considered an outlier and is eliminated. Otherwise, the congestion contrast must be applied to decide. This last restriction in the number of valid observations stands for the fact that vehicle "*l*" could also be a motorbike dodging the congestion (see Fig. 5A.2).

Congestion Contrast

In case the overtaking contrast is not meaningful and "$t_{i,j,k}^{(p)}$" remains as a doubtful observation, the congestion contrast has the last word. As there does not exist a vehicle that has overtaken the vehicle "*k*" in a valid travel time, showing the evidence of the possibility of travelling within the validity window, the only possibility to decide if the observation results from a voluntary stop or from congestion in the itinerary is by estimating the traffic state within every highway section. If there is a high probability of congestion within some stretch of the itinerary, the observation is considered as valid. In contrast, if free flowing conditions are estimated, then the observation is considered an outlier.

Toll tickets are not the best data source to decide whether there is congestion or not in the highway in real time, due to the MTT nature of these data. Loop detector data would be more suitable for this objective (e.g. assessing the occupation at each loop detector), opening a gap for data fusion schemes. However, from toll tickets the origin destination matrix can be obtained on a reconstructed basis (i.e. once the vehicles have left the highway). Therefore, an approximation to the flow in each section can be obtained for the previous time interval (Daganzo 1997). Then, in the congestion contrast it is assumed that the probability of congestion in the present time interval depends on the traffic flow in the previous time interval. If this flow exceeds 75 % of the capacity of the infrastructure, a high probability of congestion is given to the highway section, and the observation is considered valid.

Highway capacities can be estimated using the HCM (Transportation Research Board 2000), but note that the capacity of an infrastructure is a dynamic variable and should be modified in case of an incident or bad weather conditions. Then, an accurate application of the data filtering process requires some kind of information input to modify default capacities in incident conditions. In addition, not only main trunk capacities must be assessed, also the exit toll gate capacities, as congestion can arise in the off-ramp due to limited exit capacity. These capacities can be easily obtained by taking into account the number of open gates and the type of these gates (i.e. manual payment—230 veh/h, automatic credit card payment—250 veh/h and non-stop ETC systems—700 veh/h).

5A.3.2 Deciding on Type 2 Doubtful Observations

An itinerary travel time observation falling in the lower doubtful zone can result from two situations:

- A motorbike dodging the traffic jam by wriggling between cars. This type of observation should be considered as an outlier.

Appendix 5A: Obtaining a Representative Average ... 147

- There is a sudden travel time decrease in the highway due to congestion dissipation. Then the observation should be considered as valid.

This type of outlier is not as problematic as the ones in excess, for two reasons. Firstly, there are few outliers of this nature. And secondly, in congestion dissipation, traffic flows at capacity, and it is not usual to find itineraries with few observations. On this basis, the doubtful observation is considered to be a motorbike if the difference between its travel time and the other existing valid observations in the set is considered large.

More formally, assuming there exists at least one valid observation in the set of observations where the doubtful "$t_{i,j,k}^{(p)}$" is contained, and vehicle "l" is the vehicle whose itinerary travel time, "$t_{i,j,l}^{(p)}$," is the minimum within the valid observations, then "$t_{i,j,k}^{(p)}$" is considered a motorbike if:

$$t_{i,j,k}^{(p)} \leq t_{i,j,l}^{(p)} - \gamma \cdot IQR_{i,j}^{(p-1)} \qquad (5A.7)$$

In case Eq. (5A.7) holds, the observation is only considered as an outlier and eliminated if the number of originally valid observations in the set exceeds "$N_{i,j}^{(p)}/2$". Otherwise, the observation is considered as valid (see Fig. 5A.3).

5A.3.3 Estimation of the Interquartile Range in Sets with Few Data

Once the data filtering process has decided on the validity of the observations in the set, and "$\hat{t}_{i,j}^{(p)}$" has been calculated as the median of the valid observations, the only remaining variable to estimate in order to set the confidence intervals for the next time interval (using Eqs. 5A.2 and 5A.4) is the interquartile range "$I\hat{Q}R_{i,j}^{(p)}$". In the considered set with few data, the statistical calculations of "$\hat{q}_{3\,i,j}^{(p)}$" and "$\hat{q}_{1\,i,j}^{(p)}$" are not meaningful. Therefore, it is assumed that the interquartile range does not vary from the last time interval. Under this assumption, the 25 % and the 75 % percentiles of the travel time distribution for the considered "p" time interval can be calculated as:

$$\begin{aligned} \hat{q}_{1\,i,j}^{(p)} &= \hat{t}_{i,j}^{(p)} - (t_{i,j}^{(p-1)} - q_{1\,i,j}^{(p-1)}) \\ \hat{q}_{3\,i,j}^{(p)} &= \hat{t}_{i,j}^{(p)} + (q_{3\,i,j}^{(p-1)} - t_{i,j}^{(p-1)}) \end{aligned} \qquad (5A.8)$$

5A.4 Sets with No Data

It can happen, particularly at nighttime, that there is no travel time observation for a particular itinerary ("$N_{i,j}^{(p)} = 0$"). Note that in this case "$\alpha_{i,j}^{(p)} = 0$". The process to

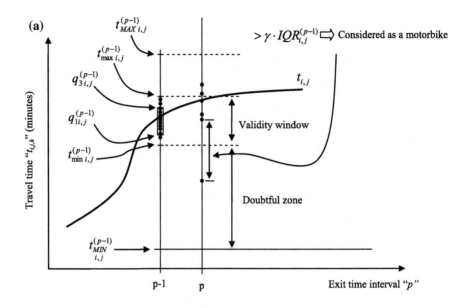

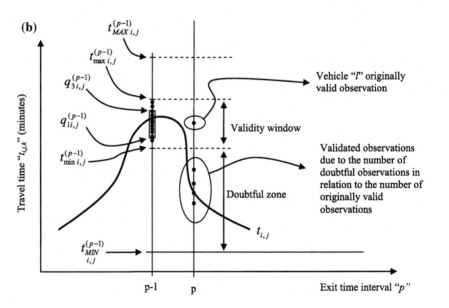

Fig. 5A.3 Deciding on lower travel time doubtful observations. **a** Elimination of an outlier. **b** Validation of observations in a congestion dissipation situation

obtain "$t_{i,j}^{(p)}$" in this situation is simple. Travel times are assumed to maintain a linear constant evolution from the two last time intervals. This assumption is considered valid if at least one of these last time intervals has some observations. Otherwise, travel time is set to a default free flow travel time. The resulting formulation for sets with no data is:

$$t_{i,j}^{(p)} = \begin{cases} t_{i,j}^{(p-1)} + \left(\frac{\alpha_{i,j}^{(p-1)} + \alpha_{i,j}^{(p-2)}}{2}\right) \cdot \left(t_{i,j}^{(p-1)} - t_{i,j}^{(p-2)}\right) & \text{if } \alpha_{i,j}^{(p-1)} > 0 \quad \text{or} \quad \alpha_{i,j}^{(p-2)} > 0 \\ t_{i,j,0} & \text{o.w.} \end{cases} \quad (5A.9)$$

In relation to the interquartile range, if "$\alpha_{i,j}^{(p-1)} > 0$" or "$\alpha_{i,j}^{(p-2)} > 0$" then it is calculated in the same way as in Sect. A.3.3. Otherwise the interquartile range is set to the default value "$IQR_{i,j,0}$".

Taking into account the fact of the related values of "0" for the reliability indicator "$\alpha_{i,j}^{(p)}$" of the measurement, does not give any specific weight to these measurements in the smoothing equation for the next time intervals.

Appendix 5B: Accurate Formulation of the Basic Algorithm

Figure 5B.1 represents a zoom of the "$i, i+1$" section of Fig. 5.2. From this figure it can be seen that Eq. (5.4) and by extension Eqs. (5.1), (5.3), (5.6), (5.7), (5.9), (5.11), (5.21), and (5.29) are not accurate.

To be accurate, Eq. (5.6) should be rewritten as:

$$t_{ex(i+1)} = t_{i,i+1} - t_{s-cut(i,i+1)} - t_{en(i)} \quad (5B.1)$$

where the single section trimmed travel time can be obtained as:

$$t_{s-cut(i,i+1)} = \frac{t_{s(i,i+1)} \cdot l_{s-cut(i,i+1)}}{l_{s(i,i+1)}} \quad (5B.2)$$

with "$l_{s(i,i+1)}$" and "$l_{s-cut(i,i+1)}$" the section length and the trimmed section length respectively.

Applying the same modification to the two-section travel time, Eq. (5.11) should be rewritten as:

$$t_{ex2(i+2)} = t_{i,i+2} - t_{s-cut(i,i+2)} - t_{en(i)} \quad (5B.3)$$

$$t_{s-cut(i,i+2)} = \frac{t_{s(i,i+2)} \cdot \left(l_{s(i,i+1)} + l_{s-cut(i+1,i+2)}\right)}{l_{s(i,i+1)} + l_{s(i+1,i+2)}} \quad (5B.4)$$

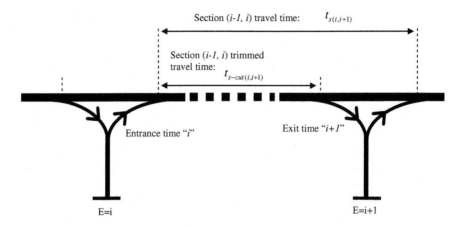

Fig. 5B.1 Detailed sketch of a highway junction

These modifications, which in practice have little consequence due to the magnitude of the section length in relation to the inner junction highway length, can also be applied to Eqs. (5.3), (5.9), (5.21) and (5.29).

This trimmed single section travel time can also be applied in the reconstruction of vehicle itineraries. Note that when reconstructing a trajectory by adding single section travel times, the last section travel time to consider must be a trimmed one.

Appendix 5C: Fusion of Different Estimations of Single Section Travel Times

From the basic and extended algorithm, different estimations of the single section travel times are obtained. Specifically, in an off-line application of the algorithm the following can be obtained:

- "$m - (i+1)$" first order estimations of single section travel times "$t_{s(i,i+1)}^{(p)}$"
- "$m - (i+2)$" second order estimations of two section travel times "$t_{s(i,i+2)}^{(p)}$", which result in "$2 \cdot (m - (i+2))$" single section second order travel times "$t_{s2(i,i+1)}^{(p)}$".

In case of real time application, "m", the ordinal number identifying the last exit in the highway should be replaced by its limitation "$i + n_i^{*(p)}$" (see Sect. 4.1).

Appendix 5C: Fusion of Different Estimations …

5C.1 Fusion of First Order Single Section Travel Time Estimations

The objective of this data fusion is to obtain a representative average of the different estimations of a particular single section travel time. To do so, a simple weighted average is applied. The two aspects to consider in the determination of the weighting factor for each one of the estimations are:

- As the length of the itineraries used to estimate a single section travel time increase, the reliability of the estimation decreases. This accounts for the greater variance of the travel time distribution in longer itineraries, and for the decrease of the trajectories overlapping zone (see Sect. 4.1).
- An accuracy indicator of each single section travel time estimation can be obtained as the minimum of the statistical significances in the travel time itineraries that take part in the calculation, "$\min\left(\alpha_{i,j}^{(p)}, \alpha_{i+1,j}^{(p)}\right)$".

Under these assumptions, the fused first order single section travel time is obtained as[3]:

$$t_{s1(i,i+1)}^{(p)} = \frac{\sum_{j=i+2}^{m}\left\{\left[\frac{1+\min\left\{\alpha_{i,j}^{(q)}, \alpha_{i+1,j}^{(q)}\right\}}{\ln(j-i)}\right] \cdot \left[\left(t_{i,j}^{(q)} - t_{en(i)}\right) - \left(t_{i+1,j}^{(q)} - t_{en(i+1)}\right)\right]\right\}}{\sum_{j=1+2}^{m}\left[\frac{1+\min\left\{\alpha_{i,j}^{(q)}, \alpha_{i+1,j}^{(q)}\right\}}{\ln(j-i)}\right]} \quad (5C.1)$$

Recall from Sect. 4.2 that "q" equals "p" for the real time application.

Again, the real time application version of Eq. (5C.1) is obtained by simply replacing "$i + n_i^{*(p)}$" instead of "m" and "p" instead of "q".[4]

Taking into account the concepts detailed in Sect. 3.1 and Appendix 5B, the first order exit time is obtained as:

$$t_{ex1(i+1)}^{(p)} = t_{i,i+1}^{(p)} - t_{s1-cut(i,i+1)}^{(p)} - t_{en(i)}^{(p)} \quad (5C.2)$$

---

[3]Note the change in notation in Appendix 5C n relation to Sect. 5.3.1 of the main text. "$t_{s1(i,i+1)}^{(p)}$" stands for the fused first order single section travel time. Likewise, "$t_{s2(i,i+1)}^{(p)}$" stands for the fused single section travel time coming from the second order algorithm. Finally "$t_{s(i,i+1)}^{(p)}$" is the notation for the fused first and second order single section travel time. This is the information to be disseminated. Same notation criteria applies to the exit times. Section 5.3.1 notation is maintained for the clarity of the concepts.

[4]This modification applies for all the remaining equations of this 5.ix. The default equations are presented in its off-line version.

Finally, an accuracy indicator of these fused first order estimations is defined in Eq. (5C.3). These accuracy indicators only consider the first two estimations of the single section travel time, which are obtained from the shortest itineraries.

$$\alpha_{s1(i,i+1)}^{(p)} = \max\left\{\min\left\{\alpha_{(i,i+2)}^{(q)}, \alpha_{(i+1,i+2)}^{(q)}\right\}, \min\left\{\alpha_{(i,i+3)}^{(q)}, \alpha_{(i+1,i+3)}^{(q)}\right\}\right\}$$
$$\alpha_{ex1(i+1)}^{(p)} = \min\left\{\alpha_{(i,i+1)}^{(p)}, \alpha_{s1(i,i+1)}^{(p)}\right\}$$

(5C.3)

5C.2 Fusion of Second Order Single Section Travel Time Estimations

Proceeding in the same way and under the same assumptions of the previous section, the second order fused two-section travel time is obtained as:

$$t_{s2(i,i+2)}^{(p)} = \frac{\sum_{j=i+3}^{m} \left\{\left[\frac{1+\min\{\alpha_{i,j}^{(q)}, \alpha_{i+2,j}^{(q)}\}}{\ln(j-i)}\right] \cdot \left[(t_{i,j}^{(q)} - t_{en(i)}) - (t_{i+2,j}^{(q)} - t_{en(i+2)})\right]\right\}}{\sum_{j=i+3}^{m} \left[\frac{1+\min\{\alpha_{i,j}^{(q)}, \alpha_{i+2,j}^{(q)}\}}{\ln(j-i)}\right]}$$

(5C.4)

To obtain the related fused second order single section travel time, recall Eqs. (5.13), (5.14), (5.15) and (5.16):

$$\left.\begin{array}{l} t_{s2(i,i+1)}^{(p)} = t_{s2(i,i+2)}^{(p)} - t_{s1(i+1,i+2)}^{(p)} \\ t_{s2(i+1,i+2)}^{(p)} = t_{s2(i,i+2)}^{(p)} - t_{s2(i,i+1)}^{(p)} \end{array}\right\} \quad \text{if } \alpha_{s1(i+1,i+2)}^{(p)} \geq \alpha_{s1(i,i+1)}^{(p)} > 0$$

$$\left.\begin{array}{l} t_{s2(i+1,i+2)}^{(p)} = t_{s2(i,i+2)}^{(p)} - t_{s1(i,i+1)}^{(p)} \\ t_{s2(i,i+1)}^{(p)} = t_{s2(i,i+2)}^{(p)} - t_{s2(i+1,i+2)}^{(p)} \end{array}\right\} \quad \text{if } \alpha_{s1(i,i+1)}^{(p)} > \alpha_{s1(i+1,i+2)}^{(p)} > 0 \quad (5\text{C}.5)$$

$$\left.\begin{array}{l} t_{s2(i,i+1)}^{(p)} = \frac{t_{s2(i,i+2)}^{(p)} \cdot l_{s(i,i+1)}}{l_{s(i,i+2)}} \\ t_{s2(i+1,i+2)}^{(p)} = \frac{t_{s2(i,i+2)}^{(p)} \cdot l_{s(i+1,i+2)}}{l_{s(i,i+2)}} \end{array}\right\} \quad \text{if } \alpha_{s1(i,i+1)}^{(p)} = \alpha_{s1(i+1,i+2)}^{(p)} = 0$$

Finally, an accuracy indicator of these fused second order estimations of the single section travel times can also be defined as:

$$\alpha_{s2(i,i+1)}^{(p)} = \alpha_{s2(i+1,i+2)}^{(p)} = \min\left\{\alpha_{(i,i+3)}^{(p)}, \alpha_{(i+2,i+3)}^{(p)}, \max\left\{\alpha_{s1(i,i+1)}^{(p)}, \alpha_{s1(i+1,i+2)}^{(p)}\right\}\right\}$$

$$\alpha_{ex2(i+2)}^{(p)} = \min\left\{\alpha_{(i,i+2)}^{(p)}, \alpha_{s2(i+1,i+2)}^{(p)}\right\}$$

(5C.6)

5C.3 Fusion of First and Second Order Single Section Travel Times

To obtain the final estimation of the single section travel time "$t_{s(i,i+1)}^{(p)}$", it is only necessary to calculate a weighted average of first and second order estimations. The weighting factors are the accuracy indicators of each one of the estimations.

$$t_{s(i,i+1)}^{(p)} = \frac{\alpha_{s1(i,i+1)}^{(p)} \cdot t_{s1(i,i+1)}^{(p)} + \alpha_{s2(i,i+1)^+}^{(p)} \cdot t_{s2(i,i+1)^+}^{(p)} + \alpha_{s2(i,i+1)^-}^{(p)} \cdot t_{s2(i,i+1)^-}^{(p)}}{\alpha_{s1(i,i+1)}^{(p)} + \alpha_{s2(i,i+1)^+}^{(p)} + \alpha_{s2(i,i+1)^-}^{(p)}}$$

(5C.7)

Where the superscripts (+) and (−) in the second order estimations refer to the two possibilities of obtaining a single section travel time from a two-section travel time (i.e. "$t_{s2(i,i+1)}^{(p)}$" can be obtained from "$t_{s1(i,i+2)}^{(p)}$" and from "$t_{s1(i-1,i+1)}^{(p)}$").
Applying a similar weighted average to the exit times:

$$t_{ex(i+1)}^{(p)} = \frac{\alpha_{ex1(i+1)}^{(p)} \cdot t_{ex1(i+1)}^{(p)} + \alpha_{ex2(i+1)}^{(p)} \cdot t_{ex2(i+1)}^{(p)}}{\alpha_{ex1(i+1)}^{(p)} + \alpha_{ex2(i+1)}^{(p)}}$$

(5C.8)

Equations (5C.7) and (5C.8) are valid for "$i \in (0, \ldots, m-1)$" (see Fig. 5.2). Note that for "$i = 0$" and for "$i = m - 1$", some of the terms in Eqs. (5C.7) and (5C.8) are not defined (i.e. "$t_{s2(i,i+1)^+}^{(p)}$" is not defined for "$i = m - 1$", "$t_{s2(i,i+1)^-}^{(p)}$" and "$t_{ex2(i+1)}^{(p)}$" are not defined for "$i = 0$"). In these situations, the related accuracy indicators have a value of zero, and Eqs. (5C.7) and (5C.8) still hold.

References

Abdulhai, B., & Tabib, S. M. (2003). Spatio-temporal inductance-pattern recognition for vehicle reidentification. *Transportation Research Part C, 11*(3–4), 223–239.
Cassidy, M. J., & Ahn, S. (2005). Driver turn-taking behavior in congested freeway merges. *Transportation Research Record, 1934*, 140–147.

Coifman, B. (2001). Improved velocity estimation using single loop detectors. *Transportation Research Part A, 35*(10), 863–880.

Coifman, B. (2002). Estimating travel times and vehicle trajectories on freeways using dual loop detectors. *Transportation Research Part A, 36*(4), 351–364.

Coifman, B., & Cassidy, M. (2002). Vehicle reidentification and travel time measurement on congested freeways. *Transportation Research Part A, 36*(10), 899–917.

Coifman, B., & Ergueta, E. (2003). Improved vehicle reidentification and travel time measurement on congested freeways. *ASCE Journal of Transportation Engineering, 129*(5), 475–483.

Coifman, B., & Krishnamurthya, S. (2007). Vehicle reidentification and travel time measurement across freeway junctions using the existing detector infrastructure. *Transportation Research Part C, 15*(3), 135–153.

Cortés, C. E., Lavanya, R., Oh, J., & Jayakrishnan, R. (2002). General-purpose methodology for estimating link travel time with multiple-point detection of traffic. *Transportation Research Record, 1802*, 181–189.

Daganzo, C. F. (1997). *Fundamentals of transportation and traffic operations*.Pergamon: Elsevier.

Dailey, D. A. (1999). Statistical algorithm for estimating speed from single loop volume and occupancy measurements. *Transportation Research Part B, 33*(5), 313–322.

Davies, P., Hill, C., & Emmott, N. (1989). Automatic vehicle identification to support driver information systems. In *Proceedings of IEEE Vehicle Navigation and Information Systems Conference*, 11–13 September, A31–35.

Dion, F., & Rakha, H. (2006). Estimating dynamic roadway travel times using automatic vehicle identification data for low sampling rates. *Transportation Research Part B, 40*(9), 745–766.

Herrera, J. C., Work, D. B., Herring, R., Ban, X., Jacobson, Q., & Bayen, A. M. (2010). Evaluation of traffic data obtained via GPS-enabled mobile phones: the Mobile Century field experiment. *Transportation Research Part C, 18*(4), 568–583.

Hopkin, J., Crawford, D., & Catling, I. (2001). *Travel time estimation*. Summary of the European Workshop organized by the SERTI project, Avignon, November 2001.

Li, R., Rose, G., & Sarvi, M. (2007). Using automatic vehicle identification data togain insight into travel time variability and its causes. *Transportation ResearchRecord* 1945, 24–32.

Li, R., Rose, G., & Sarvi, M. (2006). Evaluation of speed-based travel time estimation models. *ASCE Journal of Transportation Engineering, 132*(7), 540–547.

Lucas, D. E., Mirchandani, P. B., & Verma, N. (2004). Online travel time estimation without vehicle identification. *Transportation Research Record, 1867*, 193–201.

Mikhalkin, B., Payne, H., & Isaksen, L. (1972). Estimation of speed from presence detectors. *Highway Research Record, 388*, 73–83.

Muñoz, J. C., & Daganzo, C. F. (2002). The bottleneck mechanism of a freeway diverge. *Transportation Research Part A, 36*(6), 483–505.

Nam, D. H., & Drew, D. R. (1996). Traffic dynamics: method for estimating freeway travel times in real time from flow measurements. *ASCE Journal of Transportation Engineering, 122*(3), 185–191.

Ohba, Y., Ueno, H., & Kuwahara, M. (1999). Travel time calculation method for expressway using toll collection system data. In *Proceedings of the 2nd IEEE Intelligent Transportation Systems Conference*, pp. 471–475.

Palen, J. (1997). The need for surveillance in intelligent transportation systems. *Intellimotion, 6*(1), 1–3.

Pushkar, A., Hall, F., & Acha-Daza, J. (1994). Estimation of speeds from single-loop freeway flow and occupancy data using cusp catastrophe theory model. *Transportation Research Record, 1457*, 149–157.

Sun, C., Ritchie, S. G., & Tsai, K. (1998). Algorithm development for derivation of section-related measures of traffic system performance using inductive loop detectors. *Transportation Research Record, 1643*, 171–180.

Sun, C., Ritchie, S. G., Tsai, K., & Jayakrishnan, R. (1999). Use of vehicle signature analysis and lexicographic optimization for vehicle reidentication on freeways. *Transportation Research Part C, 7*(4), 167–185.

References

SwRi, (1998). *Automatic vehicle identification model deployment initiative—System design document*. Report prepared for TransGuide, Texas Department of Transportation, Southwest Research Institute, San Antonio, TX.

Transportation Research Board. (2000). *Highway capacity manual*, Special Report209, National Research Council, Washington, DC.

Turner, S. M., Eisele, W. L., Benz, R. J., & Holdener, D. J. (1998). *Travel time data collection handBook*. Research Report FHWA-PL-98-035. Federal Highway Administration, Office of Highway Information Management, Washington, DC.

van Arem, B., van der Vlist, M. J. M., Muste, M. R., & Smulders, S. A. (1997). Travel time estimation in the GERIDIEN project. *International Journal of Forecasting, 13*(1), 73–85.

van Lint, J. W. C., & van der Zijpp, N. J. (2003). Improving a travel-time estimation algorithm by using dual loop detectors. *Transportation Research Record, 1855*, 41–48.

Chapter 6
Short-Term Prediction of Highway Travel Time Using Multiple Data Sources

Abstract The development of new traffic monitoring systems and the increasing interest of road operators and researchers in obtaining reliable travel time measurements, motivated by society's demands, have led to the development of multiple travel time data sources and estimation algorithms. This situation provides a perfect context for the implementation of data fusion methodologies to obtain the maximum accuracy from the combination of the available data. This chapter presents a new and simple approach for the short term prediction of highway travel times, which represent an accurate estimation of the expected travel time for a driver commencing on a particular route. The algorithm is based on the fusion of different types of data that come from different sources (inductive loop detectors and toll tickets) and from different calculation algorithms. Although the data fusion algorithm presented herein is applied to these particular sources of data, it could easily be generalized to other equivalent types of data. The objective of the proposed data fusion process is to obtain a fused value more reliable and accurate than any of the individual estimations. The methodology overcomes some of the limitations of travel time estimation algorithms based on unique data sources, as the limited spatial coverage of the algorithms based on spot measurement or the information delay of direct travel time itinerary measurements when disseminating the information to the drivers in real time. The results obtained in the application of the methodology on the AP-7 highway, near Barcelona in Spain, are found to be reasonable and accurate.

Keywords Travel time estimation · Data fusion · Loop detectors · Toll ticket data · Bayesian combination · The accuracy

6.1 Introduction

Most developed countries, unable to carry on with the strategy of expanding transportation infrastructures once they become saturated, are now focusing their efforts on the optimization of infrastructure usage by means of operational and management improvements. This policy results from environmental, budget and land occupancy limitations, the latter being especially restrictive in metropolitan areas where high population density is combined with increasing mobility needs of society.

The availability of accurate and reliable travel time information appears to be the key factor for an improved management of road networks, since it allows an effective estimation of traffic states and provides the most valuable and understandable information for road users (Palen 1997). This evidence has not gone unnoticed by some European countries (Spain, France, Denmark, Italy, Finland, United Kingdom, Sweden, the Netherlands, Norway and Germany) grouped under the Trans-European Road Network (TERN), which are currently developing travel time estimation projects (Hopkin et al. 2001).

This interest expressed by transportation agencies and highway operators in addition to the development of ITS (Intelligent Transportation Systems) has led to a new framework in traffic data management and has increased the variety of reliable, precise and economically viable road surveillance technologies (Klein 2001; Skesz 2001; Martin et al. 2003; US DOT 2006). In addition, the appearance of ATIS (Advanced Traveler Information Systems) has made possible a simple and efficient dissemination of information addressed to the road user.

This context results in a new situation where it isn't rare that for a particular highway stretch to have several traffic measurements coming from different available sources (primarily in congested metropolitan highways). In addition, different algorithms or methodologies have been developed to obtain travel time estimations from these traffic measurements, obtaining a remarkable accuracy, not without an extensive research effort in recent years. Turner et al. (1998) gives a comprehensive overview of these travel time estimation methods. However, each of these methods is flawed by the intrinsic characteristics of the original measurement. The availability of different travel time estimations from different data sources usually results in complementary flaws of the estimations. This opens up new horizons for data fusion techniques applied to different estimations of road trip travel time.

Researchers' interest in travel time data fusion techniques has been increasing since the late 90s. In the USA Palacharla and Nelson (1999) studied the application of fuzzy logic to travel time estimation, evaluating which hybrid system was more effective (i.e. the fuzzy based on a neural network or on an expert system). They conclude that the neural network hybrid system is more precise, increasing the quality of the results obtained with classical travel time estimation methods. A similar methodology was tested in 2006 by the Austrian Department of Traffic, Innovation and Technology (Quendler et al. 2006), whose objective was to obtain reliable travel times and to determine the congestion level of the road network using

6.1 Introduction

multiple data sources (inductive loop detectors, laser sensors and floating taxi cars). A reduction of 50 % in the number of mistakes in the estimation of traffic state is claimed using this methodology known as ANFIS (Adaptable Neural Fuzzy Inference System). Later, Sazi-Murat (2006) relied on ANFIS to obtain delay times in signalized intersections, achieving better results than the Highway Capacity Manual (TRB 2000), mainly in heavily congested situations. Also Lin et al. (2004) and Tserekis (2006) deal with the short-term prediction of travel time in arterials, decomposing the total delay into link delay and intersection delay. The authors propose a simple model and prove a reasonable degree of accuracy under various traffic conditions and signal coordination levels.

Researchers in Singapore and China have tried to obtain predictions of traffic stream states using Bayesian inferences on a neural structure from a unique source of data. The results improve those using simple neural networks in 85 % of the situations (Weizhong et al. 2006). Park and Lee (2004), both Koreans, have obtained travel time estimations in urban areas by implementing neural networks and Bayesian inferences, both independently, and using data from inductive loops and floating cars. In both cases the results are considered promising.

In France, researchers have developed conceptually simple data fusion techniques. The best examples are the works of El Faouzi (2005a, b) and El Faouzi and Simon (2000) in the evidential Dempster-Shafer inference, which could be considered a generalization of Bayesian theory, improving the results of classical Bayes theories in pilot test runs on a highway near Toulouse. Two sources of data were used: license plate matching and inductive loop detectors. These experiences are being used by French highway operators for the estimation of travel times in their corridors (Ferré 2005; AREA 2006; Guiol and Schwab 2006).

Swedish and Scottish road operators (SRA—Sweden Road Administration and Transport for Scotland) have since 2001 been analyzing the implantation of data fusion systems to obtain road travel times in their networks. The Scottish pilot test on the A1 motorway in the surroundings of Edinburgh uses up to 4 data sources: tracking of cellular phones, inductive loop detectors, floating car data and license plate matching. Surprisingly, the cellular phone tracking, despite its lack of location accuracy in dense urban environments, stands out for its reliability (Peterson 2006; Scott 2006).

In the Netherlands, van Lint et al. (2005) use neural networks for the prediction of travel times with gaps in the data, obtaining satisfactory results in spite of this partial information. Recently, van Hinsbergen and van Lint (2008) propose a Bayesian combination of travel time short-term prediction models in order to improve the accuracy of the predictions for real time applications of this information. Results are promising, but further research is recommended increasing the number and diversity of the models to combine.

In this context, the present chapter proposes a new data fusion approach for travel time estimation in order to provide real time information to drivers entering a highway. Note that to accomplish this objective, not only accurate measurement is necessary, but also short term prediction of travel times (Rice and van Zwet 2001). A two level fusion process is proposed. On the one hand, the first fusion level tries to overcome the spatial limitations of point measurements, obtaining a representative estimation for the whole stretch. On the other hand, the second level of fusion tries not only to measure but also to predict travel times in order to achieve a more reliable estimation for the real time dissemination of the information.

This methodology results in a simple travel time data fusion algorithm, which when implemented on top of existing data collection systems, allows us to exploit all the available data sources and outperforms the two most commonly used travel time estimation algorithms. The results of a pilot test on the AP-7 highway in Spain are outlined in the chapter and show that the developed methodology is sound.

The chapter is organized as follows: Sect. 6.2 describes the different natures of travel time measurements and highlights the main objective and contribution of the chapter. In Sect. 6.3, the basic and simple algorithms used to obtain the source travel time data to fuse are presented. Section 6.4 describes the methodology used for the development of the two level data fusion system. Section 6.5 presents the results of the application of the model from the AP-7 highway in Spain. Finally, some general conclusions and issues for further research are discussed in Sect. 6.6.

6.2 Travel Time Definitions Revisited

There are two main methodologies used to measure travel time on a road link: the direct measurement and the indirect estimation. The direct travel time measurement relies on the identification of a particular vehicle on two points of the highway. These control points define the travel time target stretch. By simply cross-checking the entry and exit times of the identified vehicles, the travel time from one point to the other is obtained. The data collection techniques used in this first approach are defined by the identifying technology. Identification by means of license plates or using the toll tags ID (i.e. on-board electronic devices to pay the toll at turnpikes equipped with electronic toll collection—ETC—systems) are commonly used. These technologies are grouped under the AVI (Automated Vehicle Identification) systems. Usual problems in this direct travel time measurement are obtaining a representative number of identifications [this problem is particularly severe when the identification tag given to the vehicles has a low market penetration; take as an example the case of the TransGuide system in San Antonio, Texas, where the AVI tags were given solely for the purpose of estimating travel times (SwRI 1998)] or the elimination of frequent outliers (e.g. a driver stops for a break). In addition, and probably the most important shortcoming of direct measurements for real time applications, is the information delay. Travel time measurements are obtained once the vehicle has finished its itinerary. Therefore, a direct travel time measurement results in what will be named an MTT (i.e. Measured Travel Time) and represents a measurement of a past situation involving a delay in the real time dissemination of information (equal to the travel time). Obviously this drawback turns out to be more severe as travel time increases (i.e. long itineraries or congestion episodes).

6.2 Travel Time Definitions Revisited

The alternative is the indirect travel time estimation. It consists of measuring any traffic flow characteristic (usually the fundamental variables—flow, density and speed) on some particular points of the target stretch, and applying some algorithm in order to obtain the travel time estimation. Under this configuration, every portion of the length of the stretch is assigned to one of the measurement spots, and it is assumed that the point measurement reflects the homogeneous traffic conditions of the whole portion. Obviously the main problem with these types of methods is the lack of fulfilment of this last assumption, mainly in heavy traffic conditions when traffic conditions can vary dramatically within the assigned portion of highway. To limit the effects of this drawback, a high density of detection sites are necessary, reducing the length of highway assigned to each detector. Other limitations of these methods are the lack of accuracy of the detection technologies used to measure traffic variables, considering the inductive loop detectors as the most widely used.

In this indirect travel time estimation, the itinerary travel time (i.e. the total travel time in the target stretch) is obtained by the addition of the travel times in the portions of highway that configure the stretch. This total itinerary measure, named as ITT (i.e. Instantaneous Travel Time) in the present chapter, uses only the last available data and assumes that traffic conditions will remain constant in each section indefinitely. The main advantage of ITT is the immediacy in travel time data, reflecting the very last events on the highway. Usually, this "last minute" information is considered as the best estimation of future traffic evolution. This leads ITT estimation to be used as a naïve approach to the predicted travel time (PTT), which represents an estimation of the expected travel time for a driver entering the highway at a particular instant.

Note that in fact the ITT is a virtual measurement in the sense that no driver has followed a trajectory from which this travel time results. Particularly interesting are the works of van Lint and van der Zijpp (2003) and Li et al. (2006) where the authors propose and evaluate different algorithms in order to reconstruct the real vehicle trajectories from only ITT measurements in order to obtain an estimation of the MTT. The key question in this trajectory offline reconstruction (i.e. the information delay does not matter, only accuracy, as real time information is not the objective) is how the spot speeds are generalized over space. Constant, linear or smoothed approximations are evaluated in the referenced chapters.

By way of illustration, Fig. 6.1 shows an example of the implications of different trip travel time constructions. The information delay in the case of trip MTT involves very negative effects in the case of dramatic changes in traffic conditions during this time lag (e.g. an incident happens). The construction of trip ITT by means of section travel times reduces this information delay and the resulting travel time inaccuracies. As the traffic conditions do not remain constant until the next single section travel time update, the trip ITT also differs from the true PTT.

Figure 6.1 only aims to show the benefits of ITT immediacy in travel time prediction and the drawbacks of the MTT delayed information. In practice, the usual lack of accuracy of section travel times added up to obtain the itinerary ITT, can spoil all the benefits, resulting in situations where MTT would be just as good/bad as ITT. In addition it is also possible that some particular evolutions of

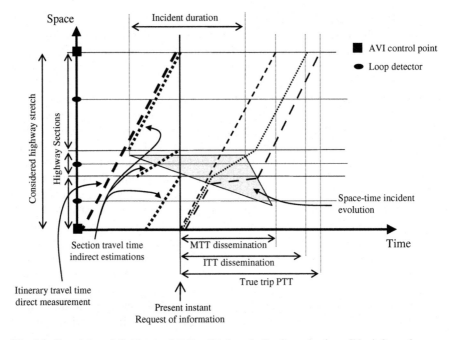

Fig. 6.1 Travel time definitions and its implications in the dissemination of the information

traffic state result in paradoxical situations where, although an accurate ITT estimation being available, this does not imply being a better approach to PTT in relation to MTT.

6.3 Naïve Travel Time Estimation Algorithms

Although not being the objective of this chapter, it is necessary for the complete understanding of the process to describe how the original data to fuse are obtained. These original data consist of two different ITT estimations and one MTT measurement. It is worthwhile to highlight the fact that the algorithms to obtain these first travel time estimations and the technologies that provide the source measurements presented in this section are in no way limiting, and could be substituted for any other procedure or technology that provide the same type of result (i.e. two ITT estimations and one MTT measurement).

In this context, this section presents an algorithm meant to obtain travel times from spot speed measurements (obtained from dual loop detectors), an algorithm to estimate travel times from loop detector traffic counts and a procedure to obtain travel time measurements from toll ticket data. The accuracy of these algorithms is not the objective of the research. Therefore, the proposed algorithms only explore

6.3.1 Spot Speed Algorithm for Travel Time Estimation

As stated earlier, this method is based on the speed measurement on a highway spot by means of dual electromagnetic loop detectors. Travel time could then be obtained by simply applying the following equation:

$$T_{1(i,t)} = \frac{l_i}{\bar{v}_{(i,t)}} \qquad (6.1)$$

where:

$T_{1(i,t)}$ is the average travel time in the highway section "i" and time interval $(t-1, t)$, obtained from algorithm 1 (spot speed algorithm—1st ITT estimation)

l_i is the length of section "i", the portion of the highway stretch considered to be associated with loop detector "i" (see Fig. 6.2)

$\bar{v}_{(i,t)}$ is the spatial mean speed measured in loop detector "i" and time interval $(t-1, t)$, calculated as the harmonic mean of the "$n_{(i,t)}$" individual vehicles' speed "$v_{k(i,t)}$" measured during $(t-1, t)$

$$\bar{v}_{(i,t)} = \frac{n_{(i,t)}}{\sum_{k=1}^{n_{(i,t)}} \frac{1}{v_{k(i,t)}}} \qquad (6.2)$$

The hypothesis considered in the application of this algorithm is that traffic flow characteristics stay constant on the whole stretch and throughout the whole time period. To limit the dramatic effects of this first source of error, a high density of surveillance and a frequent actualization of variables is needed. As an order of magnitude, this algorithm provides reasonably accurate estimations by itself when there is one loop detector every 500 m, and the updating interval is less than 5 min.

In addition, in congestion situations with frequent stop-and-go traffic, the measured spatial mean speed can be very different from the real mean speed of traffic flow, as detectors only measure the speed of the vehicles when they are moving, and do not account for the time that vehicles are completely stopped. As a result of this flaw, travel time estimations using this algorithm in congested situations can be largely underestimated. To overcome this problem, different smoothing schemes could be applied. For example averaging the measured speed in loop detector "i" and time interval $(t-1, t)$ with previously measured speeds in time and space. These evolutions of the algorithm are not considered here, where the lack of

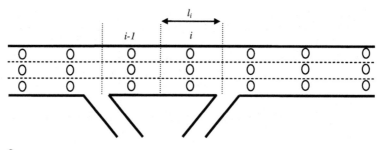

Fig. 6.2 Required surveillance configuration to apply the spot speed algorithm

accuracy of spatial mean speed in congested situations is taken into account increasing the margin of error of this travel time estimation.

Finally, in the real-world application of this type of algorithm, usually a third source of error arises. This results from the fact that the common practice at Traffic Management Centres is to compute the time mean speed of traffic stream (i.e. arithmetic average of individual vehicle speeds) instead of the space mean speed (i.e. harmonic average detailed in Eq. 6.2), the one that relates distance with average travel time (Eq. 6.1). A local time mean speed structurally over estimates the space mean because faster observations are overrepresented (Daganzo 1997). Therefore average travel times computed in this situation will be slightly underestimated. Space mean speeds are considered to be available in the rest of the chapter (Fig. 6.2).

6.3.1.1 Expected Error of the Spot Speed Algorithm

As will be described in the next section, an important element in the first fusion level is the expected error of each one of the individual values to fuse. Hence, it is necessary to define the expected error of the ITT algorithms.

As stated before, the expected error in the spot speed algorithm for the travel time estimation arises mainly from two reasons: the spatial generalization of a point measurement and the lack of accuracy of this point measurement. The magnitude of the first source of error is estimated by altering the assigned portion of highway of each detector to the most unfavorable situation. In addition, it is considered that a loop detector can overestimate the mean speed of traffic in 100 % in stop&go situations (i.e. half the time moving and half the time stopped). Stop&go traffic is considered likely to occur when the measured mean speed at the detector site falls below 80 km/h. Equations (6.3)–(6.5) together with the sketch in Fig. 6.3 describe this margin of error.

6.3 Naïve Travel Time Estimation Algorithms

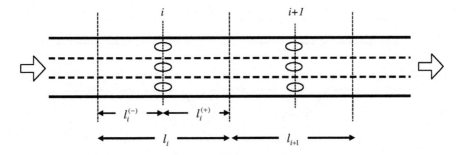

Fig. 6.3 Determining the expected error of the spot speed algorithm

$$T^{min}_{1(i,t)} = \frac{l_i^{(-)}}{\max(\bar{v}_{(i-1,t)}, \bar{v}_{(i,t)})} + \frac{l_i^{(+)}}{\max(\bar{v}_{(i,t)}, \bar{v}_{(i+1,t)})} \quad (6.3)$$

$$T^{max}_{1(i,t)} = \frac{l_i^{(-)}}{\min(\bar{v}^{min}_{(i-1,t)}, \bar{v}^{min}_{(i,t)})} + \frac{l_i^{(+)}}{\min(\bar{v}^{min}_{(i,t)}, \bar{v}^{min}_{(i+1,t)})} \quad (6.4)$$

where:

$$\bar{v}^{min}_{(i,t)} = \begin{cases} \bar{v}_{(i,t)} & \text{if } \bar{v}_{(i,t)} \geq 80 \text{ km/h} \\ 0.5 \cdot \bar{v}_{(i,t)} & \text{if } \bar{v}_{(i,t)} < 80 \text{ km/h} \end{cases} \quad (6.5)$$

The error incurred in case of using time mean speeds (instead of space mean speeds) in congested conditions, is included in the "measurement" error of loop detectors in stop&go situations. In free flowing conditions, this situation implies a slight underestimation of travel times (approximately 2 %), which can be considered negligible in relation to the spot speed generalization error.

6.3.2 Cumulative Flow Balance Algorithm for Travel Time Estimation

An alternative exists for estimating travel times from loop detector data in highway stretches that do not benefit from the required surveillance density needed for the application of the spot speed algorithm. The cumulative flow balance algorithm estimates travel time directly from loop detector traffic counts, without the previous calculation of speed. This solves the problem of the lack of accuracy in the mean speed estimation in congested situations, and allows the usage of single loop detectors with the same accuracy as the double loop in the traffic counts, but they are unable to accurately estimate vehicle speeds. The algorithm uses the entrance

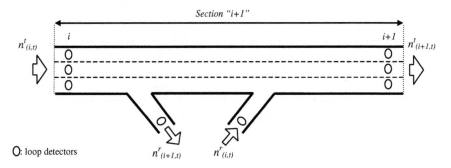

Fig. 6.4 Required surveillance configuration to apply the cumulative flow balance algorithm

and exit flows in the highway stretch in order to calculate the travel time by using a simple flow balance method.

To apply this algorithm, all the highway ramps must be equipped with loop detector units. The surveillance scheme required is displayed in Fig. 6.4, where "$n^t_{(i,t)}$" is the input traffic count of the main highway trunk in the time interval ($t - 1, t$), "$n^t_{(i+1,t)}$" is the trunk output count and "$n^r_{(i,t)}$" and "$n^r_{(i+1,t)}$" are the entering and exiting counts through the ramps comprised between detectors "i" and "$i + 1$".

Then the total entering and exiting flows are:

$$n_{(i,t)} = n^t_{(i,t)} + n^r_{(i,t)}$$
$$n_{(i+1,t)} = n^t_{(i+1,t)} + n^r_{(i+1,t)}$$
(6.6)

The travel time estimation from these traffic counts is depicted in the N-curve diagram (cumulative counts vs. time) sketched in Fig. 6.5, where:

$N_{(i,t)}$ is the cumulative traffic count that have entered the highway section at time "t"

$N_{(i+1,t)}$ is the cumulative traffic count that have exited the highway section at time "t"

$$N_{(i+1,t)} = \sum_{j=1}^{t} n_{(i+1,j)}$$
(6.7)

$S_{(i+1, t)}$ is the vehicles accumulation in the Section "$i + 1$" of the highway at time "t"
$t_{n(i+1,t)}$ is the travel time of the vehicle counting the number "n" at detector "$i + 1$", that exits at time "t"

The average travel time in the time interval ($t - 1, t$) can be estimated by calculating the shadowed area in Fig. 6.5 and dividing it by the number of vehicles that have exited the section within this time interval.

6.3 Naïve Travel Time Estimation Algorithms

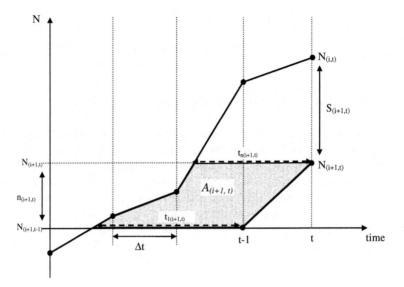

Fig. 6.5 N-Curve diagram depicting the cumulative flow balance algorithm

$$T_{2(i+1,t)} = \frac{A_{(i+1,t)}}{n_{(i+1,t)}} \quad (6.8)$$

The subscript "2" in "$T_{2(i+1,t)}$" refers to the fact that this average travel time has been obtained using algorithm 2 (cumulative flow balance algorithm—2nd ITT estimation).

Again, this is one of the simplest formulations of the algorithm, and several modifications can be applied, for instance to take into account the actual vehicle accumulation at time "t" ($S_{(i+1,t)}$). A literature review of these methods is given in Nam and Drew (1996).

The main problem with this type of formulations is the lack of exact accuracy in the detector counts, or specifically, the error in the relative counts of two consecutive detectors (i.e. all the vehicles entering the section must exit given enough time, and therefore the differences in cumulative counts between consecutive detectors should tend to zero when flows reduce to almost zero). This phenomenon is known as loop detector drift, and its effects are greatly magnified when using relative cumulative counts in consecutive detectors, due to the accumulation of the systematic drift with time. A simple drift correction has been applied to account for this fact, requiring that all the vehicles entering the section during a complete day must have exited on the same day. The drift correction factor is formulated as:

$$\delta_{(i+1)} = \frac{N_{(i,24h)}}{N_{(i+1,24h)}} \quad (6.9)$$

Then the corrected count at detector "i + 1" is equal to:

$$N_{(i+1,t)} = N_{(i+1,t-1)} + \delta_{(i+1)} \cdot n_{(i+1,t)} \tag{6.10}$$

6.3.2.1 Expected Error of the Cumulative Count Algorithm

The main source of error in this travel time estimation algorithm is the detector drift. Despite using a "δ" factor to account for this flaw, this correction is based on "historic" measurements of the two detectors and considers the average detector drift over an entire day, which do not have to correspond exactly to the current drift of the detectors in a particular traffic state. Variations of ±0.5 % in this drift correction factor should not be considered as rare. Then, the margin of error of the travel time estimations using this algorithm can be formulated as:

$$N_{(i+1,t)}^{\min} = N_{(i+1,t-1)} + (\delta_{(i+1)} - 0.005) \cdot n_{(i+1,t)} \tag{6.11}$$

$$N_{(i+1,t)}^{\max} = N_{(i+1,t-1)} + (\delta_{(i+1)} + 0.005) \cdot n_{(i+1,t)} \tag{6.12}$$

Finally,

$$T_{2(i+1,t)}^{\max} = \frac{A_{(i+1,t)}^{\max} \quad (\text{use } N_{(i+1,t)}^{\min})}{n_{(i+1,t)} \quad (\text{use } N_{(i+1,t)}^{\min})} \tag{6.13}$$

$$T_{2(i+1,t)}^{\min} = \frac{A_{(i+1,t)}^{\min} \quad (\text{use } N_{(i+1,t)}^{\max})}{n_{(i+1,t)} \quad (\text{use } N_{(i+1,t)}^{\max})} \tag{6.14}$$

6.3.3 Travel Time Estimation from Toll Ticket Data

Travel time data can be directly obtained by measuring the time taken for vehicles to travel between two points on the network. On toll highways, the data needed for the fee collection system, can also be used for travel time measurement, obtaining an MTT.

On a highway with a "closed" tolling system, the fee that a particular driver has to pay at the toll plaza varies depending on his itinerary (origin-destination). In contrast, in an "open" highway system, toll plazas are strategically located so that all drivers pay the same average fee at the toll gate. In a closed highway system, each vehicle entering the highway receives a ticket (real—usually a card with magnetic band—or virtual—using an ETC device-), which is collected at the exit. The ticket includes the entry point, and the exact time of entry. By cross-checking entry and exit data, the precise time taken by the vehicle to travel along the itinerary

6.3 Naïve Travel Time Estimation Algorithms

(route) can be determined. A similar procedure can be applied in a highway with an open toll system, by identifying the vehicle at two consecutive payment sites.

Averages can be obtained from the measurements for all the vehicles travelling along the same itinerary in the network during a time interval $(t-1, t)$. For each particular vehicle "k" travelling along a highway, the travel time spent on its itinerary between "i" (origin) and "j" (destination) expressed as "$T_{i,j,k,t}$" can be obtained by matching the entry and exit information recorded on its toll ticket. The average travel time for the itinerary in a particular time interval can be obtained by averaging the travel times of all vehicles that have exited the highway within this time period and have travelled along the same itinerary "(i, j)".

$$T_{3(i,j,t)} = median\ (T_{i,j,k,t}) \quad \forall\ k\ exiting\ the\ stretch\ (i,j),\ during\ (t-1,t) \quad (6.15)$$

where:

$T_{i,j,k,t}$ is the travel time for the itinerary "i, j" for a particular vehicle "k" that has exited the highway stretch within the time interval $(t-1, t)$

$T_{3(i,j,t)}$ is the average travel time for the itinerary "i, j" in a particular time period $(t-1, t)$. Subscript "3" stands for the usage of algorithm 3: travel time from toll ticket data. This is an MTT estimation

Note that the median is considered instead of the arithmetic mean to exclude the negative effects of outliers, which would result in an overestimation of the travel times. From these calculations, the MTT (measured travel times) are obtained.

The error in these measurements appear in the case of a great fraction of outliers (e.g. vehicles stopping for a break while travelling through the target stretch) in relation to the total identified vehicles. This situation only occurs when the total number of identified vehicles in the time interval is low. This can happen during late night hours or in the case of frequent updating of the information. This is the reason why the updating time interval of MTT is usually not lower than 15 min. In this situation, the error in these measurements can be omitted as its magnitude is much lower than the one obtained from point estimations. A detailed description of the travel time estimation process using toll ticket data can be found in Chap. 5.

6.4 Data Fusion Methodology

The travel time data fusion process proposed in this chapter is a two level fusion.

In a first level, the two ITT estimations are fused. As stated before, one of these indirect travel time measurements is obtained from speed measurements, and the other from traffic counts. Both measurements could be obtained from loop detectors. The first fusion tries to overcome the main limitations of these travel time estimation algorithms, which are on the one hand the spatial coverage limitations of the spot speed algorithm and on the other hand the lack of exact accuracy of traffic counts.

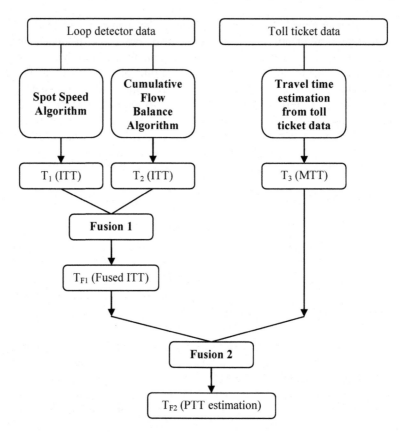

Fig. 6.6 Structure of the data fusion process

In the second fusion level, the resulting fused ITT from the first fusion is fused to an MTT, in the present application obtained from toll ticket data, which could also be obtained by any other identifying technology. The objective of this second fusion level is to increase the predictive capabilities of the accurately fused ITT by means of the information provided by the MTT, in order to obtain a better approximation to the true PTT. The data fusion structure can be seen in Fig. 6.6.

6.4.1 First Level Data Fusion

The first data fusion process fuses the two instantaneous travel times "T_1" (from a spot speed algorithm) and "T_2" (from a cumulative flow balance algorithm) to obtain "T_{F1}" a more accurate and uniform ITT in terms of spatial and temporal coverage. The objective of this first fusion is to reduce the flaws affecting each individual estimation using a more accurate fused estimation. Recall that these

6.4 Data Fusion Methodology

flaws are the spatial generalization and the accuracy of the speed measurements in the spot speed algorithm and the detector count drift in the cumulative flow balance algorithm.

The proposed fusion operator is a context dependent operator with constant mean behaviour (Bloch 1996). Since it is context dependent, it is necessary to define three contexts A, B and C. In each context the data fusion algorithm will follow a slightly different expression. Their definitions are:

- **A context**:

$$T^{\max}_{p(i,t)} \geq T^{\max}_{q(i,t)} \text{ and } T^{\min}_{p(i,t)} \leq T^{\min}_{q(i,t)} \tag{6.16}$$

where p, q refers to the estimation algorithm $p, q = 1, 2$ $p \neq q$

- **B context**:

$$T^{\max}_{p(i,t)} \geq T^{\max}_{q(i,t)} \text{ and } T^{\min}_{p(i,t)} \geq T^{\min}_{q(i,t)} \text{ and } T^{\min}_{p(i,t)} \leq T^{\max}_{q(i,t)} \tag{6.17}$$

where p, q refers to the estimation algorithm $p, q = 1, 2$ $p \neq q$

- **C context**:

$$T^{\min}_{p(i,t)} \geq T^{\max}_{q(i,t)} \tag{6.18}$$

where p, q refers to the estimation algorithm $p, q = 1, 2$ $p \neq q$

Finally, the first level fused travel times are obtained applying the following fusion operator:

$$T_{F1(i,t)} = \begin{cases} T_{q(i,t)} & \text{if } A \\ \dfrac{T^{\max}_{q(i,t)} + T^{\min}_{p(i,t)}}{2} & \text{if } B \text{ or } C \end{cases} \tag{6.19}$$

The analytic expression for the margin of error in this case is:

$$T^{\max}_{F1(i,t)} = \begin{cases} T^{\max}_{q(i,t)} & \text{if } A \text{ or } B \\ T^{\min}_{p(i,t)} & \text{if } C \end{cases} \tag{6.20}$$

$$T^{\min}_{F1(i,t)} = \begin{cases} T^{\min}_{q(i,t)} & \text{if } A \\ T^{\min}_{p(i,t)} & \text{if } B \\ T^{\max}_{q(i,t)} & \text{if } C \end{cases} \tag{6.21}$$

In Fig. 6.7 it is possible to observe that the operator is consistent and that its behaviour is clearly determined by the context. In the A and B contexts, the resulting error is smaller or equal to the smallest of the errors "ε_1" and "ε_2" (defined as the difference between the maximum and the minimum possible travel times) while in context C this could not be true. Context C must always be avoided since it

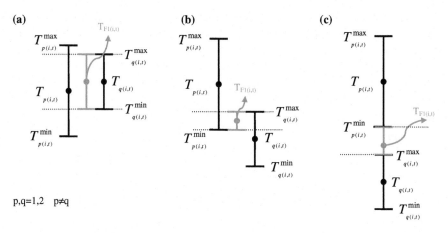

Fig. 6.7 First level ITT fusion contexts

represents flawed behaviour of the algorithm in the determination of "T_1", "T_2" or "ε_1" and "ε_2". This means that each original estimation algorithm needs a minimum accuracy and a correct estimation of the margin of error for the correct functioning of the fusion algorithm.

6.4.2 Second Level Data Fusion

The second data fusion process starts with the fused ITT (T_{F1}) and the MTT from toll tickets (T_3). On the one hand, "T_{F1}" can be considered an accurate real time measurement of the travel time in the target stretch, and could be seen as a picture of the travel time in the highway stretch reflecting the traffic state in the ($t-1, t$) time interval. Recall that this complete picture is composed of several partial pictures, each one representing a portion of the stretch (Fig. 6.1). The complete picture is obtained by simply joining these partial ones. On the other hand, the MTT, (T_3), results from direct travel time measurements, and due to its direct measurement nature it can also be considered as an accurate measurement (especially if the time interval lasts long enough to obtain a representative median).

The problem with these accurate travel time measurements is the time lag that exists between them and in relation to the final objective of the estimation: the predicted travel time (PTT) that will take a vehicle just entering the highway stretch. "T_{F1}" (ITT) is a real time picture (but nothing ensures that the traffic will remain constant during the next time interval), while "T_3" (MTT) results from the trajectories of vehicles that have recently travelled through the stretch, and therefore it is a measurement of a past situation, particularly in long itineraries. In this context the objective of the second fusion level is to gain knowledge of PTT from the two outdated accurate measurements.

6.4.2.1 Spatial and Temporal Alignment

Unlike in fusion 1, in the fusion 2 process the information is not provided by the same data source, and hence, the data will not be equally located in space and time. Therefore, a spatial and temporal alignment is needed before the data can be fused.

Usually the distance between AVI control points (the toll tag readers in this case) is greater than the spacing of loop detectors. In this situation, the spatial alignment of the measurements is directly given by the construction of the ITT from smaller section travel times (obtained from the fusion of point measurements) that match the MTT target stretch. This chapter assumes this context and that the target highway stretch to estimate travel times fits with the stretch limited by these AVI control points (see Fig. 6.1). In practice, these assumptions represent a common situation.

In relation to the temporal alignment, generally the updating frequency of ITT measurements is higher than the MTT one. This results from the fact that the accuracy of MTTs increases with the duration of the time intervals considered (which are the inverse of the updating frequency). The chosen dimension for the temporal alignment is the smallest time interval (corresponding to ITT), while the MTT is maintained constant until the next update (MTT remains constant during several ITT updates, due to its lower updating frequency). These concepts are described in more detail in Fig. 6.8 and Table 6.1.

6.4.2.2 Second Level Fusion Operator

Once ITT (T_{F1}) and MTT (T_3) are spatially and temporally aligned it is possible to apply the second fusion algorithm. This operator uses the probabilistic logic (Bloch 1996), based on Bayes' Theory:

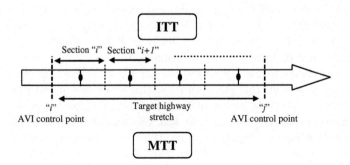

Fig. 6.8 Spatial alignment

Table 6.1 Original data for the spatial and temporal alignment

			Travel time data estimations					Spatial and temporal aligned data to fuse
			Space					
Time			ITT	Section "i"	Section "$i+1$"	Section "$i+2$"	Section "$i+3$"	
	MTT	ITT	MTT	Stretch (i,j)				
		Δt	ITT	$T_{F1(i,t)}$	$T_{F1(i+1,t)}$	$T_{F1(i+2,t)}$	$T_{F1(i+3,t)}$	$T_{F1(i,t)} + T_{F1(i+1,t)} + T_{F1(i+2,t)} + T_{F1(i+3,t)}$
			MTT	$T_{3(i,j,T)}$				$T_{3(i,j,T)}$
	ΔT	Δt	ITT	$T_{F1(i,t+1)}$	$T_{F1(i+1,t+1)}$	$T_{F1(i+2,t+1)}$	$T_{F1(i+3,t+1)}$	$T_{F1(i,t+1)} + T_{F1(i+1,t+1)} + T_{F1(i+2,t+1)} + T_{F1(i+3,t+1)}$
			MTT					$T_{3(i,j,T)}$
		Δt	ITT	$T_{F1(i,t+2)}$	$T_{F1(i+1,t+2)}$	$T_{F1(i+2,t+2)}$	$T_{F1(i+3,t+2)}$	$T_{F1(i,t+2)} + T_{F1(i+1,t+2)} + T_{F1(i+2,t+2)} + T_{F1(i+3,t+2)}$
			MTT					$T_{3(i,j,T)}$
		Δt	ITT	$T_{F1(i,t+3)}$	$T_{F1(i+1,t+3)}$	$T_{F1(i+2,t+3)}$	$T_{F1(i+3,t+3)}$	$T_{F1(i,t+3)} + T_{F1(i+1,t+3)} + T_{F1(i+2,t+3)} + T_{F1(i+3,t+3)}$
			MTT					$T_{3(i,j,T)}$
		Δt	ITT	$T_{F1(i,t+4)}$	$T_{F1(i+1,t+4)}$	$T_{F1(i+2,t+4)}$	$T_{F1(i+3,t+4)}$	$T_{F1(i,t+4)} + T_{F1(i+1,t+4)} + T_{F1(i+2,t+4)} + T_{F1(i+3,t+4)}$
			MTT					$T_{3(i,j,T)}$
		Δt	ITT	$T_{F1(i,t+5)}$	$T_{F1(i+1,t+5)}$	$T_{F1(i+2,t+5)}$	$T_{F1(i+3,t+5)}$	$T_{F1(i,t+5)} + T_{F1(i+1,t+5)} + T_{F1(i+2,t+5)} + T_{F1(i+3,t+5)}$
			MTT	$T_{3(i,j,T+1)}$				$T_{3(i,j,T+1)}$

Note Δt is the time interval between ITT updates, while ΔT is the time interval between MTT updates. In this table it is assumed that $\Delta T = 5 \cdot \Delta t$

$$p(PTT|ITT,MTT) = \frac{p(PTT,ITT,MTT)}{p(ITT,MTT)} = \frac{p(MTT|PTT,ITT) \cdot p(PTT,ITT)}{p(ITT,MTT)}$$
$$= \frac{p(MTT|PTT,ITT) \cdot p(ITT|PTT) \cdot p(PTT)}{p(ITT,MTT)}$$

(6.22)

As ITT and MTT are independent measurements, then results:

$$p(PTT|ITT,MTT) = \frac{p(MTT|PTT) \cdot p(ITT|PTT) \cdot p(PTT)}{p(ITT) \cdot p(MTT)} \quad (6.23)$$

where "$p(PTT|ITT,MTT)$" is the conditional probability of PTT given ITT and MTT, "$p(MTT|PTT)$" and "$p(ITT|PTT)$" are respectively the conditional probabilities of MTT and ITT given PTT and finally "$p(PTT)$", "$p(ITT)$" and "$p(MTT)$" are the individual probabilities of each estimation.

The probabilities in the right hand side of Eq. 6.23 must be obtained by a statistical analysis of the calibration samples. This means that, in order to apply this second fusion, a previous period of "learning" of the algorithm is needed. This off-line period of learning with sample data, which can be seen as a calibration of the algorithm, needs the three travel time estimations at a time to calculate all the conditional and individual probabilities. Note that obtaining the true PTT in an off-line basis is not a problem, because a PTT is only a future MTT, available in an

6.4 Data Fusion Methodology

off-line context. Hence, PTTs and MTTs are the same values but with a time lag between observations equal to the travel time.

In the determination of these probabilities, ITT and MTT are rounded up to the next whole minute, in order to obtain more representative relations. This does not affect the quality of the results, because the user perception of travel time is never lower than this minute unit.

Once the "$p(PTT|ITT, MTT)$" probabilities are determined, a maximum posterior probability decision rule is chosen. This means that given "T_{F1}" and "T_3" (the original ITT and MTT data for the second level fusion), the selected PTT, "T_{F2}" is the one which maximizes the conditional probability. The running of the fusion algorithm is very simple because after the spatial and temporal alignment, it is only necessary to check the table of probabilities and to obtain the corresponding fused PTT.

The decision to leave a result void is taken if the probability value does not overcome a threshold defined by the user of the system. This situation denotes little probability that ITT and MTT values coincide in the same section and time interval (e.g. it is slightly probable that ITT = 1 min and MTT = 15 min). A great number of voids in the running phase of the second fusion reveal great weaknesses of the original travel time estimation algorithms.

From Bayes' Theory it is also possible to obtain the accuracy of the result. Since when multiplying conditional probabilities part of the sample information gets lost, the uncertainty of the result (I) related with a pair of ITT and MTT, could be defined as:

$$I(ITT, MTT) = 1 - p(\text{PTT}|ITT, MTT) \qquad (6.24)$$

The goal of any travel time estimation system should be the reduction of this uncertainty, as this parameter is a good reliability indicator of the final result.

6.5 Application to the Ap-7 Highway in Spain

The data fusion technique proposed in this chapter was tested on the AP-7 toll highway in Spain. The AP-7 highway runs along the Mediterranean coast corridor, from the French border to the Gibraltar Strait. Nevertheless, the pilot test was restricted to the north east stretch of the highway between the "La Roca del Vallès" and the "St. Celoni" toll plazas, near Barcelona (Fig. 6.9). This stretch is approximately 17 km long.

The surveillance equipment installed on this stretch of the highway consists of 4 double loop detectors (i.e. approximately an average of 1 detector every 4 km). Moreover, the tolling system installed on the highway allows the direct measurement of travel times in the stretch. The duration of the loop detector data updating interval is 3 min, while the MTT are only updated every 15 min.

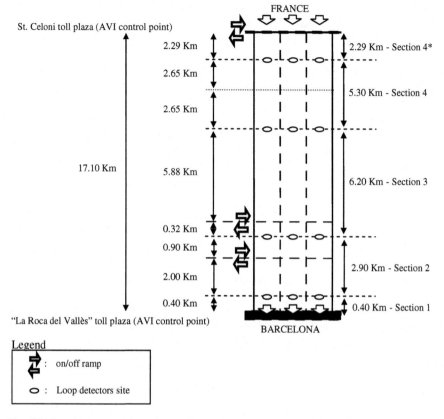

Fig. 6.9 Surveillance equipment installed on the test site

The pilot test was performed with the June 4th 2007 afternoon and evening data in the southbound direction towards Barcelona. This was a very conflictive period in terms of traffic, as it was a sunny holiday Monday in June, a time when a lot of people use this stretch of the AP-7 highway to return to Barcelona after a long weekend on the coast. The learning of the second fusion algorithm was carried out with data of a similar period from Sunday May 27th 2007.

6.5.1 First Level Fusion Results

Figure 6.10a–e shows the results of the spot speed travel time estimation algorithm "$T_{1(i,t)}$" (in the figure notation), the cumulative flow balance algorithm, "$T_{2(i,t)}$" and the results of applying the first level fusion operator, "T_{F1}", to these pairs of data in each section of the target stretch. Recall that all the information used in this level comes only from the speed and traffic count measurements at loop detector sites.

6.5 Application to the Ap-7 Highway in Spain

Note that travel time estimations for Section 4* are only available using the spot speed algorithm

To evaluate the accuracy of the fusion operator it is necessary to compare these fused travel times to the real travel times, only available for the total stretch. Note that these real travel times are in fact the final objective of the estimation (i.e. the travel time of a vehicle obtained when the vehicle is entering the highway), which could not be obtained in real time application of the algorithm, as would correspond to future information. In an offline application (like the present evaluation) these real travel times are solely the MTT (from toll ticket data) moved backwards in time a time lag equal to the experienced travel time. Only for this particular evaluation purpose, the MTT used to represent the true travel time were obtained on a three minute basis. This is shown in Fig. 6.11, where the travel time in the stretch, resulting from the reconstructed trajectories from ITT estimations in every section (from each algorithm alone and from the fused one) are compared to the real travel times in the stretch.

Several comments arise from Fig. 6.10a–e. Firstly, the inability of the spot speed algorithm to accurately describe the travel time variations resulting from the spatial evolution of jammed traffic can be clearly seen. A clear example is part (a) of Fig. 6.10. In this first section of the stretch, vehicles stop at "La Roca del Vallès" toll plaza to pay the toll fee. From 13.00 h until 21.45 h there were long queues to cross the toll gates. These queues of stopped traffic were not long enough to reach the detector site, 400 m upstream of the toll plaza. Take into account that the three lanes of the highway turn to more than 20 at the toll plaza, in order to achieve the necessary service rate and enough storage capacity near the plaza to avoid the growth of the queues blocking the on/off ramp 2.4 km upstream. This situation results in a great underestimation of travel times if using only the spot speed algorithm.

This same drawback of the spot speed algorithm cause the sharp travel time increases that can be seen in parts (b)–(d) of Fig. 6.10. Using this algorithm, travel times remain next to the free flow travel times, unaware that the congestion is growing downstream and within the assigned section of highway and obviously underestimating the travel times, until the jam reaches the detector site, when the speed falls abruptly and the travel time sharply increases. But in this situation, the algorithm considers that the whole section is jammed (when upstream of the detector traffic could be flowing freely). This results in an overestimation of travel times.

The cumulative flow balance algorithm exhibits a smoother behaviour. Recall that the problems in this case arise due to the detector drift. Although a correction for the drift is applied taking into account the historic drift between each pair of detectors, it seems that this algorithm overestimates travel times in some periods. This is due to a higher drift in some periods of the day tested in relation to the historical observations.

From Fig. 6.11 and numerical results in Table 6.2 it can be stated that great improvements were achieved with the first fusion operator with reductions of the mean estimation errors throughout the day. However, the maximum errors in a

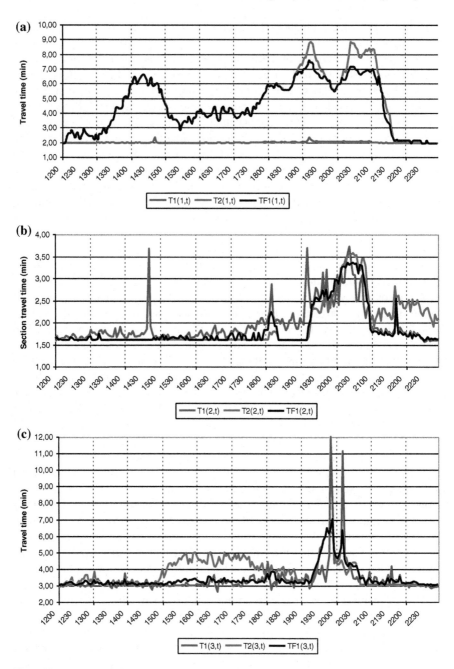

Fig. 6.10 First level fusion results on the AP-7 highway, June 4**th** 2007 data. **a** Section 1 travel times. **b** Section 2 travel times. **c** Section 3 travel times. **d** Section 4 travel times. **e** Section 6.4* travel times

6.5 Application to the Ap-7 Highway in Spain

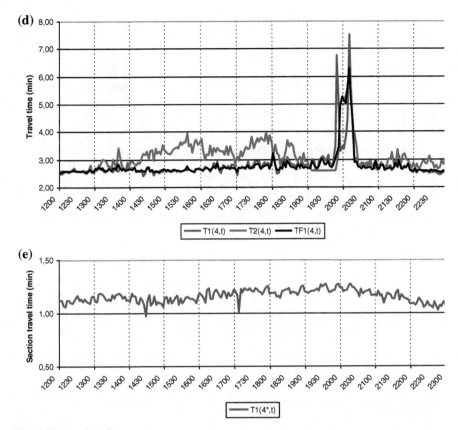

Fig. 6.10 (continued)

particular time slice slightly increased. This probably occurs due to an overestimation of the lower threshold of the spot speed algorithm margin of error, which should be lower. Note that this minimum travel time in congestion situations does not account for the possibility of congested conditions on two consecutive detector sites but traffic is free flowing in some portion of highway between them. This punctual and circumstantial increase of the maximum error does not jeopardize the great improvements achieved in global with the first fusion operator.

6.5.2 Second Level Fusion Results

Figure 6.12 shows the results of applying the second fusion operator to the test data. Recall that this fusion process considers only the whole target stretch for which MTT observations are available. The original data fused in this case are on the one hand the ITT resulting from the first fusion process "$T_{F1(i,j,t)}$", and on the other hand

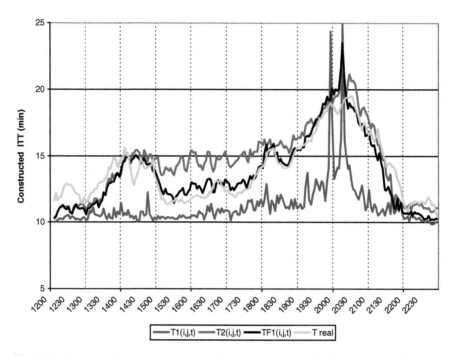

Fig. 6.11 First level fusion results on the AP-7 highway from "St. Celoni" to "La Roca del Vallès", Spain, June 4th 2007 data

Table 6.2 Accuracy of the first level of fusion

Algorithm	Mean relative error (%)	Mean absolute error (min)	Max. absolute error (min)	
			In excess	Lower
Spot speed	19.09	2.80	5.97	6.83
Cumulative	10.19	1.34	3.63	2.73
First fusion	6.01	0.81	4.60	2.93

the MTT resulting from the toll ticket data "$T_{3(i,j,t)}$", updated only every 15 min. Both sets of data are rounded to the closest whole minute.

The results of the second fusion operator "$T_{F2(i,j,t)}$" are compared with the real PTT (Predicted Travel Time), that will suffer the drivers entering the stretch at that particular time. Recall that these real PTT could not be obtained in real time application of the algorithm, as they would correspond to future information.

Again, the results of this second fusion operator are promising, and better in terms of the overall functioning than in punctual estimations (Table 6.3). The main criticism of this second level of fusion is the negative effects of the rounding to whole minutes, which causes sudden changes in the predicted travel time. This rounding is acceptable in terms of the diffusion of the information, where one

6.5 Application to the Ap-7 Highway in Spain

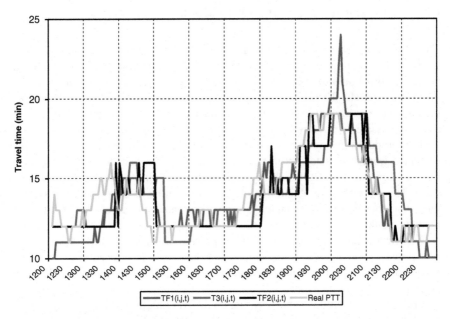

Fig. 6.12 Second level fusion results on the AP-7 highway from "St. Celoni" to "La Roca del Vallès", Spain, June 4th 2007 data

minute accuracy is normally enough. However this rounding can vary the relative differences between the data to fuse in 1 min, modifying the probabilities to consider and the corresponding result. As an issue for further research, a modified maximum posteriori probability decision rule should be analyzed. This decision rule should consist of taking into account (for instance as a weighted average) the occurrence probabilities of the two PTT values adjacent to the most likely one. This decision rule could diminish the negative effects of this necessary rounding.

Table 6.3 Accuracy of the second level data fusion

Algorithm	Mean relative error (%)	Mean absolute error (min)	Max. absolute error (min)	
			In excess	Lower
F1 ITT	7.69	1.08	6	4
Toll Ticket MTT	10.33	1.43	5	5
Second Fusion	6.83	0.97	5	4

6.6 Conclusions and Further Research

This chapter presents a simple approach for reliable road travel time estimation and short term prediction, using data fusion techniques. The objective is to obtain an accurate estimation of the travel time on a highway itinerary at the instant the driver enters the stretch. Therefore, short term forecasting is needed. The developed system can be easily put into practice with the existing infrastructure, and is able to use data obtained from any kind of sensor in any type of road link.

The proposed methodology needs several point estimations of travel times (obtained from loop detectors in the present application) and direct measurements of travel times in the target highway section (for example obtained from toll ticket data). The algorithms to obtain this original travel time data to fuse, although not the main objective of the chapter, are also presented and discussed as an intermediate result.

The fusion algorithm is a two level process using both fuzzy logic and a probabilistic approach which implements the Bayes rule. The fused travel times are found to be more reliable than the initial ones and more accurate if the learning process is carefully developed.

The results of the pilot test carried out on the AP-7 highway in Spain indicate the suitability of the data fusion system for a better usage of the different surveillance equipment already installed on the roads.

Further developments are possible with the model, for example to analyze the effects of improving the accuracy of the source data to fuse in the final estimation, or defining the requirements in the learning process to improve the probabilistic fusion.

References

AREA. (2006). Système d'information temps de parcours sur le réseau autoroutier AREA. *Asecap PULA, 2006*, 105–115.

Bloch, I., (1996). Information combination operators for data comparative review with classification. *IEEE Transactions on Systems, Man and Cybernetics—Part A* **26**(1), 52–67.

Daganzo, C. F. (1997). *Fundamentals of transportation and traffic operations*. New York: Elsevier.

El Faouzi, N. E., & Simon, C. (2000). Fusion de données pour l'estimation des temps de parcours via la théorie de l'evidence. *Recherche transports sécurité, 68*, 15–27.

El Faouzi, N. E., (2005). Bayesian and evidential approaches for traffic data fusion: Methodological issues and case study. In *Proceedings of the 85th Transportation Research Board Annual Meeting*. Washington D.C.

El Faouzi, N. E., (2005). Fusion de données en I.T. Estimation multisource du temps de parcours. *1er Séminaire TISIC*, Paris: INRETS.

Ferré, J., (2005). *Quality control and fusion of road and weather data at SAPN* [online]. Centrico Briefing Note. Available from: http://www.centrico.de/documents/briefing%20notes/20_Data_quality_control.pdf. Accessed March 18, 2007.

Guiol, R., & Schwab, N. (2006). Vers une information « temps de parcours » toujours plus performante et adaptée aux besoins de la clientèle: L'expérience ASF-ESCOTA. *Asecap PULA, 2006*, 97–104.

Hopkin, J., Crawford, D., & Catling, I. (2001). *Travel time estimation*. Avignon: Summary of the European Workshop organized by the SERTI project.

Klein, L. A. (2001). *Sensor technologies and data requirements for ITS*. Boston: Artech House Publishers.

Li, R., Rose, G., & Sarvi, M. (2006). Evaluation of speed-based travel time estimation models. *ASCE Journal of Transportation Engineering, 132*(7), 540–547.

Lin, W. H., Kulkarni, A., & Mirchandani, P. (2004). Short-term arterial travel time prediction for advanced traveller information systems. *Journal of Intelligent Transportation Systems, 8*(3), 143–154.

Martin, P. T., Feng, Y., & Wang, X., (2003). *Detector technology evaluation*. Department of Civil and Environmental Engineering, University of Utah sponsored by US Department of Transportation.

Nam, D. H., & Drew, D. R. (1996). Traffic dynamics: Method for estimating freeway travel times in real time from flow measurements. *ASCE Journal of Transportation Engineering, 122*(3), 185–191.

Palacharla, P. V., & Nelson, P. C. (1999). Application of fuzzy logic and neural networks for dynamic travel time estimation. *International Transactions in Operational Research, 6*(1), 145–160.

Palen, J., (1997). The need for surveillance in intelligent transportation systems. *Intellimotion* **6**(1), 1–3, University of California PATH, Berkeley, CA.

Park, T., & Lee, S., (2004). *A Bayesian approach for estimating link travel time on urban arterial road network*. In A. Laganà et al. (Eds.), Berlin: Springer.

Peterson, A., (2006). Travel times—A multi-purpose issue. *Proceedings of the 4th Conference of the Euro-regional Projects, I2Tern*. Barcelona.

Quendler, E., Kristler, I., Pohl, A., Veith, P., Beranek, Kubleka, H., & Boxberger, J., (2006). Fahrerassistenzsystem LISA intelligente infrastruktur. *Vernetzungsworkshop des BMVIT* **3**(I2).

Rice, J., & van Zwet, E., (2001). A simple and effective method for predicting travel times on freeways. *Proceedings of the IEEE Conference on Intelligent Transportation Systems* (pp. 227–232). Oakland, CA.

Sazi Murat, Y. (2006). Comparison of fuzzy logic and artificial neural networks approaches in vehicle delay modeling. *Transportation Research Part C, 14*(5), 316–334.

Scott, G., (2006). Scottish nacional journey system and development of data fusion. *Proceedings of the 4th Conference of the Euro-regional Projects, I2Tern*. Barcelona.

Skesz, S. L., (2001). *"State-of-the-art" report on non-traditional traffic counting methods*. Arizona Department of Transportation.

SwRI, (1998). *Automated vehicle identification tags in San Antonio: Lessons learned from the metropolitan model deployment initiative*. Science Applications International Corporation, McLean, VA.

TRB (Transportation Research Board), (2000). *Highway capacity manual*. National Research Council, Washington D.C.

Tsekeris, T. (2006). Comment on "Short-term arterial travel time prediction for advanced traveller information systems" by Wei-Hua Lin, Amit Kulkarni and Pitu Mirchandani. *Journal of Intelligent Transportation Systems, 10*(1), 41–43.

Turner, S. M., Eisele, W. L., Benz, R. J. & Holdener, D. J., (1998). *Travel time data collection handBook*. Research Report FHWA-PL-98–035. Texas Transportation Institute, Texas A&M University System, College Park.

US Department of Transportation, (2006). *T-REX project: ITS unit costs database* [online]. Research and Innovative Technology Administration (RITA). Available from: http://www.itscosts.its.dot.gov. Accessed March 18, 2007.

van Hinsbergen, C. P. I. & van Lint, J. W. C, (2008). Bayesian combination of travel time prediction models. *Proceedings of the 87th Transportation Research Board Annual Meeting.* Washington D.C.

van Lint, J. W. C., & Van der Zijpp, N. J. (2003). Improving a travel time estimation algorithm by using dual loop detectors. *Transportation Research Record, 1855,* 41–48.

van Lint, J. W. C., Hoogendoorn, S. P., & van Zuylen, H. J. (2005). Accurate freeway travel time prediction with state-space neural networks under missing data. *Transportation Research Part C, 13*(5–6), 347–369.

Weizhong, Z., Lee, D. H., & Shi, Q. (2006). Short-term freeway traffic flow prediction: Bayesian combined neural network approach. *Journal of Transportation Engineering, 132*(2), 114–121.

Chapter 7
Value of Highway Information Systems

Abstract After the generalized deployment of advanced traveler information systems, there exists an increasing concern about their profitability. The costs of such systems are clear, but the quantification of the benefits still generates debate. This chapter analyzes the value of highway travel time information systems. This is achieved by using notions of expected utility theory to develop a departure time selection and route choice model. The model assumes that every driver has a level of accepted lateness for his trip and some perceived knowledge of the travel times on the route. Only these two inputs support his decisions. The decision making process does not require the consideration of a complex cost function and does not involve any optimization. The results of the model are used to compute the unreliability costs of the trip (i.e. scheduling costs and stress) and to obtain the benefits of real-time information systems. Results show that travel time information only has a significant value when there is an important scheduled activity at the destination (e.g. morning commute trips), in case of total uncertainty about the conditions of the trip (e.g. sporadic trips), or when more than one route is possible. Systems with very high accuracy do not produce better results. The chapter also highlights the difference between the actual value that information provides to the drivers and the value they perceive, which is much smaller. For instance, massive dissemination of travel time information contributes to the reduction of day-to-day travel time variance. This favors all drivers, even those without information, although they do not realize it. This misperception suggests limited willingness to pay for travel time information.

Keywords Departure time selection · Route choice · Unreliability costs · Travel time uncertainty

7.1 Introduction and Background

For a number of years, one of the objectives of many traffic agencies around the globe has been the development of Advanced Traveler Information Systems (ATIS), and in particular the dissemination of travel time information. The boom of information and communication technologies, in short the ITS (Intelligent Transportation Systems), has opened up new possibilities, not without huge investments in surveillance and dissemination technologies (e.g. increased loop detector density, installation of Automatic Vehicle Identification systems, ..., Variable Message Signs, on board navigation devices, Smartphone applications, ...) (see Turner et al. 1998 for a comprehensive review). In general, the accuracy of travel time estimation is directly related to the intensity of surveillance and to the level of technological development of the measurement equipment, although significant research efforts have been made to improve this relationship (see Chap. 5 presenting a methodology to obtain travel times from toll ticket and without the requirement of additional surveillance equipment). In spite of this, one has to bear in mind that, in real-time information systems, some degree of information uncertainty is unavoidable, even with a perfect travel time measurement. This corresponds to the fact that real-time information, actually, is a short term prediction with intrinsic uncertainty (see Chap. 2 devoted to travel time definitions).

The costs of installing such information systems are known by traffic agencies. They explicitly pay for them. The benefits obtained, however, are not reported in such quantitative terms. It is usually claimed that travel time information is the most valuable traffic information for drivers (Palen 1997), because it allows for making better decisions (e.g. route, mode and departure time selection) and it reduces the uncertainty drivers' face. These vague statements regarding the benefits, although qualitatively true, should not be enough to justify the required investments. Furthermore, they do not help in achieving more efficient implementations. The quantification of the value of travel time information systems as a function of the system design parameters is necessary. This would allow assessing their benefit/cost ratio and defining trade-offs leading to better designs and increased system efficiency (e.g. technology selection, accuracy requirements, corridor prioritization, assessment of the dissemination strategy...).

Some studies have dealt with this quantification of benefits using stated preferences surveys (see Khattak et al. (2003) for an extensive review). In Walker and Ben-Akiva (1996) for example it is found that travelers would be willing to pay up to $0.50 per trip for convenient and accurate travel time information. Though this type of approach may give some indication of the travelers' perceived value for the information (biased, because individual drivers do not perceive the benefit to the whole group of travelers), it is totally insufficient to address the previous objectives. Much more detail is required. This is confronted in the other common approach, and adopted in the present chapter, based on the cost-benefit conceptual framework. It is considered that the value of the information is equal to the reduction in the trip costs resulting from its knowledge (Chorus et al. 2006; Ettema and Timmermans

2006; Levinson 2003). The main benefit of information is the reduction of uncertainty, which may imply significant reductions in trip costs.

Not only average delays imply costs to the driver. Travel time uncertainty (or synonymously, travel time unreliability) caused by the existence of variable day-to-day delays, also entails high costs (Bates et al. 2001; Fosgerau and Karlström 2010; Lam and Small 2001; Noland and Polak 2002; Small 1982). Uncertainty implies, simultaneously, probabilities of arriving too late and of arriving too early. Both situations result in an extension of the lost time (at destination or en-route), with the aggravating circumstance of missing meetings, connections or incurring other lateness penalties in case of arriving too late. To prevent the latter, drivers allow extra time for the journey (i.e. the buffer time), increasing the probability of arriving too early. It should be clear that as travel time uncertainty increases so does the total lost time. It is proven in (Fosgerau and Karlström 2010) that considering the simplest model to introduce risk aversion and for any given travel time distribution, the scheduling costs grow linearly with the standard deviation of travel times. Even in the unlikely situation of being exactly on time, the anxiety and stress caused by the uncertainty implies a cost for the driver. It is reported in Ettema and Timmermans (2006) that these scheduling costs due to travel time uncertainty may account for 20–40 % of the generalized trip costs. 15 % is reported in Fosgerau and Karlström (2010), and values between 5 and 85 % are found in the more detailed analysis in the present chapter (see Sect. 7.5). The costs of uncertainty, although frequently overlooked, are clear.

Unreliability should not be confused with variability. Unreliability implies uncertainty. Variability may not. On the one hand, travel time reliability can be defined as a lack of unexpected delays. Then an itinerary could be considered as reliable when the actual travel time for a particular driver is close to his expected travel time. Take into account that the expected travel time may include expected delays. On the other hand, travel time variability can be defined as the variation in travel time on the same trip traveled at different times of the day or on a different day of the week. With these definitions, travel time variability only depends on the performance of the highway, while reliability also depends on the driver's expectancy. Expected travel times vary with the driver's knowledge, which could be result of experience gained from past trips or due to the information directly provided by the road operator. Better knowledge, reduces uncertainty, and may enable drivers to make better travel decisions and reduce unreliability costs. For instance, the previous knowledge of a sporadic driver (i.e. a driver facing a low recurrence trip) is probably limited to the free flow travel time (given by the expected free flow speed for the type of road and the distance). In this case, the driver faces maximum unreliability, because the uncertainty affects the possibility of encountering delays and also their magnitude. Unreliability costs are mainly related to a lack of knowledge. On the contrary, a commuter in a freeway corridor may have a good knowledge of the existence of peak periods and their possible travel time range. This allows a better judgment in the departure time selection. In this case, unreliability costs are mainly related to freeway performance. A better management of the freeway (some possibilities are explored in Torné and Soriguera 2012) that generally accomplishes both a reduction in the average delay and in its day-to-day variability (Ettema and Timmermans 2006), would result in a reduction of the unreliability. Finally, consider a freeway with a real-time travel time information system. Drivers are informed with day specific information, which translates, for example, in a better route choice (if more than one route were available). The conclusion is clear; information reduces unreliability by reducing the stress involved in decision making. However, the value of information will be higher when it makes possible the reduction of other costs of unreliability (e.g. diminishes lateness or earliness probabilities) by modifying some trip decisions.

Two types of trip decisions could be modified. On the one hand, operative decisions, which can be made at a given instant, in real-time. Route choice and acceptance or not of park and ride options are, in general, examples of operative decisions. On the other hand, there are decisions that need to be planned in advance. Departure time choice is the most relevant planning decision. Note that in those trips where departure time choice is an option, it is difficult to figure out how a driver could decide at a particular instant to depart earlier than planned, leaving at a stroke the already planned tasks (one of those could be sleeping, for instance). Or in case of a delayed departure, would the driver be able to instantaneously fill the time gap at the origin with new productive and unplanned tasks? The most reasonable answer is no. In addition, habit, in case of commuters, implies a strong inertia in planning decisions, difficult to overcome with real-time information.

This means that in order to support a decision, information must be made available in an adequate timeline. Real-time information may support operative decisions, but it is unlikely to significantly affect planning decisions. In order to affect departure time (i.e. a planning decision) information should be given at least several hours in advance and generally much more. This would allow one to plan accordingly. Given this required long term horizon, (by far pre-trip) information must be based on historical data. Current travel times are not informative of travel times several hours later (Chrobok et al. 2004). Historical information is limited to day-to-day travel time distribution (i.e. average + variation). This type of knowledge can be gained with experience of the trip (e.g. commuters) without any kind of information system in this case. On the contrary, real-time information (that can be disseminated on-trip or just before the planned departure time—shortly pre-trip) is a short term prediction, mostly based on current travel time measurements. This type of information, always supported by an ATIS, provides daily specific travel times, accounting for possible non-recurrences. This type of information can only support operative decisions, like route choice, and will have very little effect on planning decisions, like departure time selection (see Table 7.1).

Surprisingly, this conceptual postulate seems to contradict some of the current literature on the issue (see Mahmassani and Jou (2000) for a review). It is generally claimed that travelers respond to real-time information, obtained just before the planned departure, in three possible ways: do not change anything, change the route, or change the departure time. The reported choice frequencies of each decision vary, but the departure time change option is always significant. These findings have influenced recent research on the value of travel time information

7.1 Introduction and Background

Table 7.1 Types of travel time information and supported trip decisions

Type of information	Attributes	Contents	Horizon	Supported decisions	Obtained through
Historical	Retrospective	Average + possible variation	>Several hours	Planning: departure time route choice	Experience or ATIS
Real-time	Current + predictive	Day specific value + uncertainty	<1 h	Operative: route choice departure time (limited)	ATIS

Note Adapted from Ettema and Timmermans (2006)

(Ettema and Timmermans 2006), where it is assumed that real-time information can affect departure time choice without limitations. The magnitude of this influence is clearly overlooked. Even though a significant percentage of drivers may change departure time in response to real-time quantitative information (up to 70 % in Khattak et al. 1996), it is not reasonable to think that they can act before knowing the information. Therefore, if the information is obtained 15 min before the planned departure (as in Khattak et al. 1996), this represents a maximum threshold in the amount of time that the departure can be brought forward. This maximum is difficult to achieve in practice, given the immediacy limitations of planning decisions. This issue is addressed in Mahmassani and Jou (2000), where it is found that real-time information affects departure time mostly in a magnitude between 0 and 10 min. Specifically, only around 30 % of drivers are capable of modifying more than 3 min their departure time, and given a 10 min threshold, less than 10 %. Failing to consider this limitation (like in Ettema and Timmermans 2006) clearly overestimates the potential of real-time information.

The previous discussion reflects how the value obtained from travel time information systems depend on the reduction of unreliability costs and how these costs are related to the specific characteristics of the driver, the infrastructure and the information system itself (see Table 7.2). This means that a system could be profitable in one freeway corridor and not in another. Or that, in the same corridor,

Table 7.2 Elements affecting the value of travel time information

Concerns	Element	Effect
Infrastructure	Day-to-day variability	Commuters baseline unreliability
	Existence of alternative routes	Possibility of reducing unreliability by switching routes
Driver	Experience in the corridor	Level of previous knowledge
	Importance of being on time	Magnitude of scheduling costs
Information system	Accuracy of the ATIS	Remaining unreliability

the profitability of the system may vary according to the type of drivers travelling on a particular day.

This chapter proposes a method to quantify the value of travel time information taking into account the elements shown in Table 7.2. The method simply obtains the value of the information as the difference between the trip costs with and without information. Only travel time and scheduling costs (including stress) for the trip are considered.

The method consists of a departure time and route choice model, based on the expected utility theory for scheduling trips, originally proposed in Small (1982), deriving from the seminal work in Vickrey (1969) and further developed to include travel time uncertainty (see Bates et al. (2001) or Noland and Polak (2002) for a extensive review, and Fosgerau and Karlström (2010) for a brilliant culmination. The model proposed here follows the formulation in Fosgerau and Karlström (2010), although conceptually differs in how the driver makes the decision. The traditional approach states that the driver selects the option that maximizes his expected utility (i.e. minimizing the expected trip costs, which are inversely proportional to the utility of the trip). The decision follows from a "complex" stochastic optimization where it is assumed that the driver knows his different scheduling costs (e.g. earliness penalty, lateness penalty, travel time cost) and is capable of adequately weighing the probabilities and figuring out the expectations for each option (e.g. for the different departure times). The complexity and lack of realism of this decision making process has been the object of several criticisms (Avineri and Prashker 2003; Bonsall 2003). Here, the decision is based on a much simpler process. It is only assumed that for each trip the driver accepts a probability of being late (inversely proportional to the importance of arriving on-time). This simple and realistic input supports the choice model. Under some restrictions, the results will be equivalent to those of the traditional expected utility approach, although the conceptual decision making process is much simpler. As a consequence of this conceptual framework, the model allows splitting the choice model from the costs function, and treating them independently. It also allows considering the accuracy of the information system as well as the perception error of different drivers.

The rest of the chapter is organized as follows. In Sect. 7.2 the departure time and route choice model is presented. Section 7.3 is devoted to the modeling of unreliability costs. Next, in Sect. 7.4, some of the apparent limitations of the model are discussed and some solutions proposed. In Sect. 7.5 the model is used to assess the value of travel time information systems in some typical scenarios, where information allows the limited modification of departure times and route switch (in contrast to Ettema and Timmermans (2006), where only departure time is addressed or in Levinson (2003) for route choice). Numerical results are provided. Finally, a summary and some conclusions regarding the value of highway travel time information are presented.

7.2 Modeling Departure Time and Route Choice

It is assumed that only two decisions are relevant when facing a highway trip: the departure time and the route choice. Mode shift options are not considered. It is also assumed that the differences in utility among alternative decisions are only due to differences in travel times and in scheduling costs. All other costs are neglected. Finally, it is assumed that every driver has a previous knowledge, characterized by a perceived travel time distribution, "$\tilde{\phi}_T(t)$" (the "$\sim$" superscript stands for the "perceived" attribute and "ϕ" for a probability density function). This assumption is not limiting in any sense, because this perceived distribution does not need to be similar to the real highway travel time distribution "$\phi_T(t)$". For instance, the previous knowledge of a sporadic driver is limited to the free flow travel time "t_f", and the perceived distribution would be mentally constructed from previous life experiences in similar environments. On the contrary, commuters' previous knowledge, gained with experience on the corridor would be translated into more wisely perceived distributions, where "$\tilde{\phi}_T(t) \approx \phi_T(t)$".

Because highway travel time can be thought of being composed of a minimum deterministic free flow travel time, "t_f" and a stochastic delay, the perceived travel time distribution can be formulated as:

$$\tilde{\phi}_T(t) = t_f \cdot \left(1 + \tilde{\phi}_D(d)\right) = t_f \cdot \left(1 + \tilde{\mu}_d + \tilde{\sigma}_d \cdot \tilde{\phi}_X(x)\right) \quad (7.1)$$

where "$\tilde{\phi}_D(d)$" is the perceived normalized delay distribution (d = delay/t_f, dimensionless) that can be expressed in terms of its mean, "$\tilde{\mu}_d$", standard deviation, "$\tilde{\sigma}_d$", and standardized distribution of perceived normalized delays, "$\tilde{\phi}_X(x)$" (similarly than in Fosgerau and Karlström 2010). "X" is a standardized version of the normalized delays, with mean 0 and variance 1. Because the normalized delay distribution can be roughly extrapolated between similar highway environments, with this formulation the method can be applied with the simple input of "t_f", even if travel time distributions are not available.

Given "$\tilde{\phi}_T(t)$" for the different available routes "j", the selection of the departure time and the route choice is a typical example of decision making under uncertainty. The objective of the driver will be to maximize his utility. This is achieved by minimizing the time assigned to the trip, "t_a", while fulfilling his maximum accepted lateness probability. This probability, "P_L", can be seen as the number of trips that the driver accepts to be late, over the total. For instance, in the morning commute to work, the driver could accept being late 1/10 days. This results in "$P_L = 0.1$". With these definitions, it is possible to obtain "t_a" as a function of "P_L", as:

$$\begin{aligned}
t_a = \min_j [t_{a(j)}] &= \min_j \left[\tilde{\Phi}_{T(j)}^{-1}(1-P_L) \right] \\
&= \min_j \left[t_{f(j)} \cdot \left(1 + \tilde{\Phi}_{D(j)}^{-1}(1-P_L)\right) \right] \qquad (7.2) \\
&= \min_j \left[t_{f(j)} \cdot \left(1 + \tilde{\mu}_{d(j)} + \tilde{\sigma}_{d(j)} \cdot \tilde{\Phi}_{X(j)}^{-1}(1-P_L)\right) \right]
\end{aligned}$$

where "Φ^{-1}" represents the inverse cumulative distribution function. The argument "$j*$" that minimizes Eq. 7.2 corresponds to the selected route.

The decision rule in Eq. 7.2 represents a generalization of the traditional decision making criteria under uncertainty (De Neufville 1990). For instance, the minimum expected value criterion would correspond to "$P_L \approx 0.5$" (equal only for symmetric distributions), the minimax criterion to "$P_L \approx 0$", and the minimin criterion to "$P_L \approx 1$". The method would also be equivalent to the maximization of the expected utility criterion, given an adequate relationship between "P_L" and the costs that define the utility function (e.g. lateness and earliness penalties). This relationship will be presented in more detail in the next section. However, the concept of the method does not imply the driver being aware of these costs, and does not assume that the driver makes his decision based on a complex optimization. It only assumes that he has an objective "P_L" and some previous knowledge.

The model also allows characterizing the two possible typologies of trips in relation to the departure time: the "morning" and the "evening" commute trips. On the one hand, morning commute trips, where the driver has a scheduled arrival at the destination and adapts the departure time accordingly, correspond to values of "$P_L < 0.5$". The inverse of "P_L" is a proxy for the importance of being on time. On the other hand, evening commute trips, where departure time is fixed and there is no strongly scheduled activity at destination, correspond to values of "$P_L \geq 0.5$". In this last situation, the interpretation of "P_L" is not so much the accepted probability of being late, as it is the risk level accepted in the decision making. Higher values imply risk prone decisions.

In summary, Eq. 7.2 defines the selection of the departure time and the route choice taking into account the available routes, the driver knowledge, "$\tilde{\phi}_{D(j)}(d)$" and the characteristics of the trip, included in "P_L" (Fig. 7.1).

7.2.1 Including Perception Errors

The perceived knowledge of a driver on a highway can be modeled by an average knowledge, given the driver typology (i.e. commuter or sporadic), represented by the "$\tilde{\phi}_D(d)$" distribution, plus a driver specific perception error, "ε_p". This error includes the different ways in which different drivers may perceive and remember

7.2 Modeling Departure Time and Route Choice

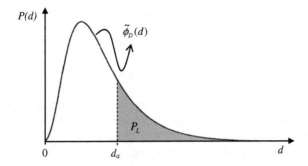

Fig. 7.1 Selection of the departure time, "d_a" (in terms of normalized delay), as a function of "P_L", without information

travel times in order to construct their mental distribution, as well as differences in their utility function not captured by travel times (different tastes, different objectives, etc....). This can be included in the model by simply considering a random term in the normalized delays term in Eq. 7.2.

$$t_a = \min_j \left[t_{f(j)} \cdot \left(1 + \tilde{\Phi}_{D(j)}(1 - P_L) + \varepsilon_p \right) \right] \quad (7.3)$$

This process for modeling the perception error is analogous to the error term included in discrete choice models with random utility (e.g. Train 2003). For instance, assuming that the error term, "ε_p" is normally distributed with 0 mean and variance "σ_p^2", Probit choice probabilities would be obtained.

7.2.2 Effects of Information

The existence of a travel time information system modifies the knowledge of the driver. By a travel time information system is meant a system that provides both historical information (by far pre-trip, for example on an annual calendar basis) and real-time information (shortly pre-trip and on-trip). Then, the perceived knowledge of every informed driver will be:

$$\tilde{\phi}_D(d) = \begin{cases} \phi_D(d) & \text{pre - trip} \\ \phi_D(d|d_i) & \text{on - trip} \end{cases} \quad (7.4)$$

where "$\phi_D(d)$" is the actual distribution of normalized delays and "$\phi_D(d|d_i)$" is the normalized delay distribution given an informed real-time delay "d_i". The characteristics of this last distribution will depend on the accuracy of the information system. By simplicity it is assumed that the system provides unbiased estimations, a desirable property, although not achieved by all systems (see Chap. 3). If biased, experience with the system would allow drivers to account for the systematic drift (Ettema and Timmermans 2006). It is also assumed that accuracy errors are normally distributed, "$\phi_D(d|d_i) \approx N(d_i, \sigma_i^2)$", and therefore:

$$\phi_D(d|d_i) = d_i + d_i \cdot c.v._{\cdot i} \cdot \phi_Z(z) \tag{7.5}$$

where "$\phi_Z(z)$" is the standard normal density function truncated at percentiles 2.5 and 97.5 % to avoid theoretically possible extreme values, and "$c.v._{\cdot i} = \sigma_i/d_i$" is the coefficient of variation of the "$\phi_D(d|d_i)$" distribution, the variable characterizing the accuracy of the information system. It is assumed that the driver is aware of this accuracy (for instance by disseminating information as a confidence interval (e.g. "$d_i - 2\sigma_i, d_i + 2\sigma_i$" for the 95 % confidence). No perception error is considered if information is available.

This new "$\tilde{\phi}_D(d)$" (from Eq. 7.4) allows modifying trip decisions, according to Eq. 7.2 (see Fig. 7.2). Information may allow a route switch, provided that several routes exist. Information may also allow changes in the departure time. For instance, sporadic drivers will be able to select their departure time, without limitations, according to the pre-trip information "$\phi_D(d)$". However, additional departure time modifications due to the real-time knowledge of "$\phi_D(d|d_i)$" will be very limited for all drivers. Recall that real-time information does not allow big changes in planning decisions. A parameter "τ" is introduced to define the maximum threshold for this departure time modification (see Table 7.3). For instance, for evening commute trips, "$\tau = 0$" because departure time is fixed.

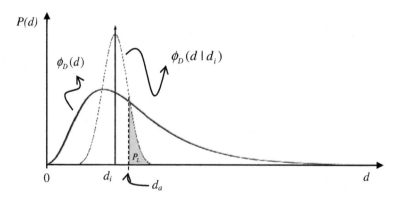

Fig. 7.2 Selection of the departure time, "d_a" (in terms of normalized delay), as a function of "P_L" with information

7.2 Modeling Departure Time and Route Choice

Table 7.3 Parameters used in the application of the model

Parameter		Value	Ref.
t_f	Free flow travel time	$t_{ff1} = t_{ff2} = 22.33$ min	25 % percentile of the off-peak travel time distribution (see Fig. 7.4)
$\phi_D(d)$	Normalized delay distribution	Constructed from empirical data on the AP-7 turnpike accessing the city of Barcelona, Spain	See Fig. 7.4 and Table 7.4. The standardized version, $\phi_X(x)$, is also presented
		Off-peak: $\mu_d = 0.14$, $\sigma_d = 0.28$	
		Peak hour: $\mu_d = 0.72$, $\sigma_d = 0.43$	
		All day: $\mu_d = 0.44$, $\sigma_d = 0.48$	
		Joint distributions: 2 routes and massive information dissemination. Obtained through simulation from previous distributions and for different accuracy levels	See Fig. 7.5
$\tilde{\phi}_D(d)$	Perceived normalized delay distribution	Commuters: $\tilde{\phi}_D(d) = \phi_D(d)$ Peak or off-peak selected accordingly	–
		Sporadic drivers: $\tilde{\phi}_D(d) = \phi_D(d)$ all day	–
ε_p	Perception error	Commuters: normal, ($\mu_p = 0$, $\sigma_p = 0.08$)	Ben-Elia and Shiftan (2010)
		Sporadic drivers: normal, ($\mu_d = 0$, $\sigma_p = 0.025$)	Jodar (2011)
ε_i	Information accuracy	Normal, ($\mu_i = 0$, $\sigma_i = c.v._i \cdot \mu_i$) Truncated at percentiles 2.5 and 97.5 % Default value: $c.v._i = 0.1$	Design variable. Sensitivity analysis performed
τ	Departure time modification threshold	Morning commute: negative exponential: $T(\tau) = 0.277 \cdot e^{-0.277\tau}$ $\bar{\tau} = 3.6$ minutes	Mahmassani and Jou (2000)
		Evening commute: 0 (zero)	

(continued)

Table 7.3 (continued)

Parameter	Value		Ref.
VOT	14.1 €/h		Asensio and Matas (2008)
$B_{(\Delta t, VOT)}$	Value of travel time differences	$B_{(\Delta t, VOT)} = \begin{cases} 0.054 \cdot VOT & \text{if } \Delta t \leq 5 \text{ min} \\ 0.615 \cdot VOT & \text{if } 5 \text{min} < \Delta t \leq 15 \text{ min} \\ VOT & \text{if } \Delta t > 15 \text{ min} \end{cases}$	AASHTO (1977)
θ_U	Maximum value of stress	1.52 €/trip	Jodar (2011)
L, e, θ_L	Lateness and earliness penalties	Morning commute	$P_L = 0.1$	$L = 4.91,$ $e = 0.61,$ $\theta_L = 11.79$	Consistent with expected utility theory and the relative relationship with Small (1982)
		Sporadic work trip	$P_L = 0.05$	$L = 3.65,$ $e = 0.61,$ $\theta_L = 8.32$	
		Evening commute and sporadic leisure trip	$P_L = 0.5$	$L = 0.70,$ $e = 0.31, \theta_L = 0$	Small (1982) for flexible trips
		Average trip	$P_L = 0.15$	$L = 2.40,$ $e = 0.61,$ $\theta_L = 5.47$	Small (1982)

7.3 Modeling Travel Time and Unreliability Costs

The objective in this section is to compute the unreliability costs of a trip given departure time and route choice decisions. The costs model proposed is based on the original formulation of the scheduling costs of unreliability (Small 1982) with some modifications to include the effects of rescheduling and stress.

The total travel time, "T", invested in the trip, is equal to:

$$T = t_a + t_L \tag{7.6}$$

where "t_a" is the time assigned to the trip and, in case of late arrival, a lateness component, "t_L", appears. If late arrival is not the case, "t_a" can be divided into the travelling part, "t_t", and some earliness "t_e".

$$t_a = t_t + t_e \tag{7.7}$$

Note that for late arrivals "$t_e = 0$" and "$t_a = t_t$". This implies that "t_e" and "t_L" cannot be different than zero simultaneously.

Both, "t_t" and "t_L" (if they exist), can also be divided into two additional components: the expected and the unexpected parts. These are defined by their relationship with "$\tilde{\mu}_t$", the perceived expected travel time "$\tilde{\mu}_t = t_f \cdot (1 + \tilde{\mu}_d) = t_f + t_f \cdot E\left[\tilde{\phi}_D(d)\right]$". Then, Eq. 7.6 can be reformulated as:

$$\begin{aligned} T &= t_t + t_e + t_L \\ &= t_{t(\exp)} + t_{t(un\exp)} + t_{L(\exp)} + t_{L(un\exp)} + t_e \end{aligned} \tag{7.8}$$

How "T" is divided among these 5 parts depends on "t" (the actual travel time), "$\tilde{\mu}_t$" and "t_a" (see Fig. 7.3). Note that expected lateness, "$t_{L(\exp)}$" would only exist in case "$\tilde{\mu}_t > t_a$". This may happen in risky evening commute decisions ("P_L" > 0.5) or in situations when real-time information modifies "$\tilde{\mu}_t$" but "t_a" cannot be adapted due to limitations in the modification of planning decisions.

Given this decomposition of "T" into different types of travel time, it is only necessary to define the corresponding cost functions for each term. This is:

$$C\_t_{t(\exp)} = VOT \cdot t_{t(\exp)} \tag{7.9}$$

where "VOT" is the average value of travel time, [monetary units/time]. No other costs apply to the part of the travel time invested in travelling and happening as expected.

$$C\_t_e = e \cdot VOT \cdot t_e \tag{7.10}$$

where "e" accounts for the earliness penalty, and represents the unproductive fraction of time in early arrivals [dimensionless]. In general "e" < 1.

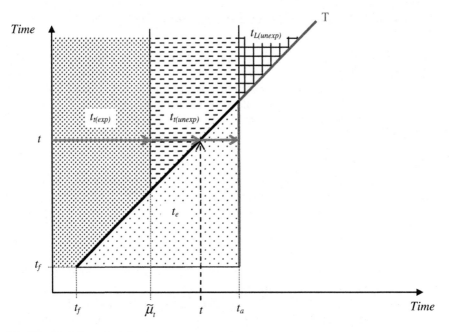

Fig. 7.3 Modeling scheduling costs

$$C\_t_{L(un\,exp)} = VOT \cdot t_{L(un\,exp)} + L \cdot VOT \cdot t_{L(un\,exp)} + \delta_{L(un\,exp)} \cdot \theta_L \cdot VOT \quad (7.11)$$

where "L" accounts for the additional unitary cost of unexpected lateness [dimensionless] and "θ_L" accounts for the fixed cost of being late [in terms of standard travel time units]. "$\delta_{L(un\,exp)}$" is a dummy variable that equals 1 if "$t_{L(un\,exp)}$" $\neq 0$, and 0 otherwise.

In case the lateness is expected, the additional costs could be reduced because the driver is given the opportunity of re-scheduling the activities and/or informing of his expected lateness. The reduction of the additional costs is assumed to be huge when the importance of being on time is relatively low, and tends quadratically to 0 as "P_L" tends to 0. Very important meetings cannot be rescheduled. This is:

$$C\_t_{L(\exp)} = VOT \cdot t_{L(\exp)} + (1 - P_L)^2 \cdot \left[L \cdot VOT \cdot t_{L(\exp)} + \delta_{L(\exp)} \cdot \theta_L \cdot VOT \right] \quad (7.12)$$

"$\delta_{L(\exp)}$" is a dummy variable that equals 1 only if "$t_{L(\exp)}$" $\neq 0$ and "$t_{L(un\,exp)}$" $= 0$, and 0 otherwise. This avoids double counting the fixed costs of lateness in case both, expected and unexpected, lateness exist.

Finally, the additional cost of consuming unexpected travel time, accommodated within the buffer, due to the stress produced by the increasing uncertainty about the arrival time is formulated as:

7.3 Modeling Travel Time and Unreliability Costs

$$C\_t_{t(unexp)} = VOT \cdot t_{t(unexp)} + (1 - P_L)^2 \cdot \left[\left(\frac{t_{t(unexp)}}{t_a - \tilde{\mu}_t}\right) \cdot \theta_U\right] \quad (7.13)$$

where "θ_U" is the maximum stress penalty [€/trip]. Equation 7.13 assumes that the stress increases linearly as the scheduled arrival time approaches, and diminishes quadratically with the increase of "P_L". Important meetings imply higher stress when consuming unexpected travel time. The adequacy of the functional relationship of Eqs. 7.12 and 7.13 in relation to "P_L", although conceptually plausible, would require further research in order to support them with empirical evidence.

From the sum of Eqs. 7.9–7.13, the total travel time unreliability cost is obtained. It can be formulated as:

$$C = \{VOT \cdot t\} + VOT \cdot \left\{ \begin{array}{l} e \cdot t_e + L \cdot t_{L(unexp)} + \delta_{L(unexp)} \cdot \theta_L \\ + (1 - P_L)^2 \cdot [L \cdot t_{L(exp)} + \delta_{L(exp)} \cdot \theta_L] \end{array} \right\} \\ + \left\{ (1 - P_L)^2 \cdot \left[\left(\frac{t_{t(unexp)}}{t_a - \tilde{\mu}_t}\right) \cdot \theta_U\right] \right\} \quad (7.14)$$

where the actual time travelling, "t", is equal to:

$$t = t_{t(exp)} + t_{t(un\,exp)} + t_{L(exp)} + t_{L(un\,exp)} \quad (7.15)$$

Note that Eq. 7.14 is grouped in three summands (between curly brackets). Each one has a different conceptual meaning. The first one is the cost of travel time. The second is the additional scheduling cost due to unreliability. And the third term is the additional cost due to stress.

7.3.1 Value of a Travel Time Information System

The value, "V", that a particular driver obtains from a travel time information system is simply the difference between the trip cost, "C", with (i) and without (wo) information.

$$V_{(t,\tilde{\mu}_t,t_a,t_i)} = C^{(wo)}_{(t,\tilde{\mu}_t,t_a)} - C^{(i)}_{(t,t_i,t_a)} \quad (7.16)$$

Equation 7.16 computes the benefits obtained from travel time differences, with and without information (i.e. "$\Delta t = t^{(wo)} - t^{(i)}$"). There is evidence in the literature (going back to AASHTO 1977) that drivers value differently the travel time variations depending on the amount of gain/loss (i.e. non-linear cost function). Drivers might be indifferent to very short changes, while increasingly penalized for longer ones. This is considered by substituting the fixed "VOT" in Eq. 7.14, by a function "$B(\Delta t, VOT)$" (for example the one proposed in Table 7.3). Because this function also affects the fixed lateness penalty, it implies some flexibility in the preferred arrival times, so that no significant disutility is incurred provided the arrival is within a "band of indifference", as empirically proven in Mahmassani and Chang (1986).

In order to obtain its average value it is needed to compute the expected value of "V":

$$E\left[V_{(t,\tilde{\mu}_t,t_a,t_i)}\right] = \int_0^\infty \left(C^{(wo)}_{(t,\tilde{\mu}_t,t_a)} - C^{(i)}_{(t,t_i,t_a)}\right) \cdot \Phi_T(t) \cdot \partial t \qquad (7.17)$$

Recall that the input variables in Eqs. 7.16 and 7.17 (i.e. "t", "$\tilde{\mu}_t$", "t_a" and "t_i") are, in fact, derived variables, obtained from the outputs of the decision model. The original inputs to the departure time and route choice model are "t_f", "$\tilde{\phi}_D(d)$", $\phi_D(d)$ "ε_D", "τ", "P_L" and "$c.v._i$".

7.3.2 Morning Commute: Equivalence with the Expected Utility Theory

The choice and the cost models are independent so far. This allows the driver making the decisions without explicitly evaluating anything related to his costs. However the costs are needed in the second step, in order to assess the decisions made and to obtain the benefit obtained from the decision support system. It is obvious that the decision, in case of morning commute trips, is related to the scheduling costs, even without the driver's explicit knowledge. Larger accepted lateness probabilities should correspond to smaller lateness penalties, and vice versa. This means that, for being consistent, it is necessary to define the link between the choice and the cost models.

This link is defined in terms of the relationship between "P_L" and the parameters of the cost function (i.e. "VOT", "e", "L", "θ_L" and "θ_U") and considering that the obtained results should be equivalent to those obtained with the standard expected utility approach. For a given set of cost parameters, there is only one "P_L" consistent with the expected utility theory. This is the one that minimizes the cost function. Or equivalently, the one for which the expected additional costs due to earliness are equal to the expected additional costs due to lateness. This relationship should be fulfilled in each application of the method. As an order of magnitude, if the cost function is simplified by neglecting "θ_L" and "θ_U", then "$P_L = \frac{e}{e+L}$" (Bates et al. 2001).

7.4 Model Limitations and Some Solutions

The implicit assumption of the model is that, given a fixed wish arrival time (morning commute) or a fixed wish departure time (evening commute), the travel time distribution for any available route does not change with the existence of information. This obviates the possibility of "concentration" and "overreaction", two possible effects that can penalize the benefits of information (Ben-Akiva et al. 1991). Departure time concentration would happen if the reduction in travel time uncertainty is translated into increased demand rates, due to more uniformity in the perception of the network traffic state. Route change overreaction may happen if, as a result of information, many drivers switch to an alternative route, taking the increase of congestion with them. In some scenarios, concentration and overreaction are not significant and can be neglected, but not in others. In this latter case, some modifications to the method are provided.

7.4.1 Limited Dissemination of the Information

Assuming that the information is only given to a small group of travelers, their decisions on departure time and route choice would not be able to change the overall pattern of travel time distributions. Hence, the assumption of the model (i.e. invariant travel time distributions) always holds, in this case. This limited dissemination of information is frequently assumed (Ettema and Timmermans 2006; Fosgerau and Karlström 2010), limiting the applicability of the proposed methods. Simulation experiments (Mahmassani and Chen 1991) found that 20 % of informed drivers are enough to contradict this assumption.

7.4.2 One Route, Massive Dissemination of Information

In this case, only departure time selection matters. It is claimed that, if the wished arrival (or departure) times do not change with information and their rates [wished arrivals/time] are approximately constant, the travel time distributions will also not change, and therefore the assumption of the model holds. Note that, given the previous conditions, the average buffer times could change from day to day, but the distribution of buffer times around the mean would be invariant (because "P_L"s are maintained). This would result in exactly the same demand/capacity rates for the infrastructure and therefore in the same travel time distributions. If we consider that most of the drivers in congested metropolitan freeways are commuters and that, in this case, real-time information modifies departure time decisions in a very narrow time window, the assumption of constant wish arrival rates in that short time interval can be accepted.

Special cases where the previous arguments do not hold exist. Consider for example annual holiday migrations, where due to the low recurrence of the episode, all drivers can be considered as sporadic. Therefore, they are able to significantly modify their departure time according to the historical information provided for the episode. In addition, the wished arrival/departure times are not fixed, because not being a daily trip, the objective is more related to avoid congestion, independently of the arrival/departure time. Given this scenario, it could happen that in periods when historical information predicts smaller travel times, huge congestion takes place due to the pulling effect (i.e. concentration). This statement has been empirically and unpleasantly tested by the author (although informally, unfortunately). It is evident that information changes travel time distributions in this case. Therefore, the results of the proposed method should not be taken literally in these very special cases. Solutions, involving the double-guessing nature of human behavior, are not treated here.

7.4.3 Two Routes, Massive Dissemination of Information

Assume two equivalent routes from an origin to a destination with a great majority of commuters. In the absence of an ATIS, their knowledge is characterized by the historical day-to-day travel time distribution on each route. From the principle of network user equilibrium (Wardrop 1952), stating that no time can be saved by switching routes, and also considering the imperfect knowledge due to uncertainty and perception errors, it can be concluded that the stochastic user equilibrium reached will result in equal travel time distributions on both routes (Emmerink et al. 1995). This result does not imply equilibrium on any particular day, because day specific information is not available. Therefore, travel times on each route are independent variables, drawn from the same distribution.

The situation changes in case of real-time information. Fully informed drivers would be able to achieve the user equilibrium every day, only blurred by the lack of accuracy of the information system. Both routes would be able to pool their capacities to serve jointly the existing demand. Given the independent fluctuations of supply for each route, the possibility of pooling the resources allows a reduction of the service time variance. This can be intuitively seen in the case of non-recurrent incidents on one route. The lack of travel time equilibrium generated would be rebalanced with the existence of an ATIS. This means that if real-time information is available, situations without travel time equilibrium will be short-lived.

From the previous discussion it is concluded that in case of two routes with massive dissemination of information, the assumption of constant travel time distribution does not hold. The implications on travel time distributions depend on the level of traffic demand, as is described next.

7.4.3.1 Effects on off-Peak Travel Time Distributions

Delays appear when the capacity of the infrastructure (i.e. the transportation supply) cannot match the demand. Both supply and demand are stochastic in nature, and this leads to uncertainty in delays, which can be characterized by the day-to-day delay distribution. During off-peak periods, average supply largely exceeds average demand. This means that only infrequent severe fluctuations in supply will produce delays (e.g. a heavy truck accident blocking the whole freeway trunk). Therefore, the day-to-day delay distribution for these periods will be sharp for null delays, with very low frequencies for a long right tail (Van Lint et al. 2008).

If two equivalent routes are available, off-peak periods can be defined as those periods when the joint capacity of both routes is always greater than the total demand, even if there is a severe capacity restriction on one of the routes. This means that the increase in demand on one route, which would happen in case of an informed incident on the alternative route, could be served without causing any additional delays.

With this assumption, the joint normalized delay distribution "$\phi_{D(pool)}(d)$" can be modeled by:

$$\phi_{D(pool)}(d) = \phi_D(\min(d_1, d_2)) \tag{7.18}$$

where "d_1" and "d_2" are the independent delays on each route, in the absence of information. For instance, if "$\phi_{D(1)}$" and "$\phi_{D(2)}$" are described by negative exponential distributions with parameters "$\lambda_{(1)}$" and "$\lambda_{(2)}$" [acceptable for off-peak periods (Van Lint et al. 2008; Li et al. 2006; Noland and Small 1995)], "$\phi_{D(pool)}$" will be equally exponentially distributed with parameter "$\lambda_{(pool)} = \lambda_{(1)} + \lambda_{(2)}$". Otherwise, in case of considering two independent, identically distributed, normal r. v.'s, with mean "μ" and variance "σ^2" [suitable for peak periods (Van Lint et al. 2008; Li et al. 2006)] it can be shown (Nadarajah and Kotz 2008) that the minimum is normally distributed with mean "$\mu - 0.56 \cdot \sigma$" and variance "$0.68 \cdot \sigma^2$". In any case, both the expected delay and its variance will be reduced in this case. Finally, take into account that the variance resulting from the information system inaccuracy must be added to the variance of the joint distribution, due to imperfect demand balance.

7.4.3.2 Effects on Peak Hour Travel Time Distributions

Peak periods are characterized by large demand/capacity ratios where slight fluctuations imply significant travel time variations. In comparison to off-peak periods, this is translated into longer average delays and a much more flat and less skewed day-to-day delay distribution (Van Lint et al. 2008).

In case of two routes, and because both routes serve vehicles near or at capacity, the rebalance of demand resulting from day specific massive information implies a travel time reduction on one route and an increase on the other. The pooling effect yields in this case a joint delay distribution with the same original expected value and half the variance (see Appendix 7A), again, plus the inaccuracy of the system.

7.5 Numerical Examples

In order to assess the value of travel time information systems, a series of numerical examples are presented in this section. 11 different scenarios are proposed (see Tables 7.3, 7.4 and 7.5 for the quantitative definition) aiming to analyze the effects of: the type of trip (i.e. morning or evening commute, sporadic trip, leisure, work,…), the traffic conditions (i.e. peak or off-peak periods), the existence of alternative routes (i.e. 1 vs. 2 available routes) and the effects of limited versus massive dissemination of information. An additional "average" scenario is proposed, with the objective to obtain an overall value of travel time information.

The parameters selected for the application of the model are presented in Tables 7.3 and 7.4 and in Figs. 7.4 and 7.5. Mostly, these have been obtained from freeways accessing the city of Barcelona, Spain. 31 days of travel time data (over the period March 3rd to October 29th, 2006) were used to construct the travel time distributions. Data was recorded from toll ticket data on a vehicle per vehicle basis on the AP-7 turnpike, between the toll plazas at "Maçanet" and "La Roca", a 44.63 km stretch (see Chap. 5 for a description of the test site and database). All these days belong to the same recurrent pattern, with clearly identifiable congested periods.

Results reported in Table 7.5 are obtained through simulation. 10,000 trips are simulated for each scenario. Values of travel time information in Table 7.5 consider a default accuracy of the information system given by "$c.v._i$" = 0.1. This means that there is a 95 % probability of an error <20 %. This accuracy level can be achieved by most ATIS implementations (see Chap. 3). The sensitivity of the results in relation to the accuracy level of the information system has been analyzed. Results from the marginal information value [in € cents] obtained from a 1 % variation of the system accuracy (in a reasonable range, $c.v._i$ = 0.01–0.5) are also reported in Table 7.5.

Table 7.4 Numerical description of standardized normalized delay distributions

	$1 - P_L$	0.95	0.90	0.85	0.80	0.75	0.70	0.65	0.60	0.55	0.5
$\Phi_X^{-1}(1 - P_L)$	Off-peak	1.40	0.81	0.49	0.31	0.18	0.09	0.01	−0.07	−0.13	−0.19
	Peak	1.82	1.34	1.01	0.75	0.51	0.32	0.16	0.02	−0.10	−0.21
	All day	1.91	1.28	0.90	0.64	0.43	0.25	0.09	−0.06	−0.19	−0.30

7.5 Numerical Examples

Table 7.5 Value of travel time information in different scenarios[a]

Scenario						Costs without info[c] [€/trip]						Value of information[c] [€/trip]								Accuracy	
#	Trip type	P_L	Traffic	# of routes	Info[b]	Travel time		Sched.		Stress		Δ travel time		Δ sched.		Δ stress		Total		$\partial V/cv_i$[c,d]	
1	Morning commute	0.1	Peak hour	1	M & L	8.54	(9.03)	2.44	(3.14)	0.01	(0.32)	0.00	(0.00)	0.07	(0.57)	0.00	(0.17)	0.21	(0.74)	0.75	(1.33)
2	Evening commute	0.5	Peak hour	1	M & L	8.52	(9.00)	0.55	(0.96)	0.00	(0.00)	0.00	(0.00)	0.00	(0.39)	0.00	(−0.06)	0.00	(0.33)	0.00	(0.43)
3			Off-Peak	1	M & L	5.71	(5.99)	0.25	(0.51)	0.00	(0.00)	0.00	(0.00)	0.00	(0.17)	0.00	(−0.07)	0.00	(0.10)	0.00	(0.26)
4[5]	Sporadic work	0.05	Peak hour	1	L	8.55	(9.04)	2.92	(7.75)	0.51	(0.59)	0.00	(0.00)	0.15	(4.33)	0.35	(0.45)	0.48	(4.78)	1.31	(1.84)
5[5]	Sporadic pleasure	0.5	Off-Peak	1	L	5.70	(5.99)	0.78	(1.00)	0.00	(0.00)	0.00	(0.00)	0.36	(0.61)	0.00	(−0.05)	0.33	(0.56)	0.49	(0.71)
6	Morning commute	0.1	Peak hour	2	L	8.55	(9.02)	2.34	(3.15)	0.08	(0.35)	0.00	(0.92)	0.03	(0.53)	0.00	(0.19)	0.36	(1.64)	1.19	(2.24)
7					M	8.74	(9.01)	1.26	(1.98)	0.15	(0.38)	0.00	(0.00)	0.02	(0.34)	−0.05	(−0.01)	0.01	(0.33)	0.73	(1.21)
8	Evening commute	0.5	Peak hour	2	L	8.53	(9.03)	0.55	(1.02)	0.00	(0.00)	0.00	(0.94)	0.00	(0.22)	0.00	(−0.08)	0.00	(1.08)	0.38	(1.40)
9					M	8.77	(9.03)	0.33	(0.61)	0.00	(0.00)	0.00	(0.00)	0.00	(0.14)	0.00	(−0.13)	0.00	(0.01)	0.63	(0.39)
10			Off-Peak	2	L	5.68	(5.96)	0.27	(0.54)	0.00	(0.00)	0.00	(0.33)	0.00	(0.06)	0.00	(−0.11)	0.00	(0.27)	0.57	(0.68)
11					M	5.26	(5.37)	0.20	(0.31)	0.00	(0.00)	0.00	(0.00)	0.00	(0.04)	0.00	(−0.15)	0.00	(−0.11)	0.73	(0.38)
12	Average	0.15	Peak hour	2	L	8.55	(9.02)	2.14	(3.17)	0.10	(0.32)	0.00	(0.92)	0.03	(0.74)	0.02	(0.17)	0.33	(1.83)	8.55	9.02
13					M	8.73	(9.01)	1.25	(2.18)	0.15	(0.34)	0.00	(0.00)	0.02	(0.56)	−0.02	(−0.01)	0.03	(0.56)	8.73	(9.01)

[a] $t_{f(1)} = t_{f(2)} = 22.33$ min, default $c.v_i = 0.1$
[b] Dissemination of information: massive versus limited
[c] Median and (Mean) values
[d] Marginal rate of substitution: value of info versus system accuracy [€ cents/1 %]. Median and (Mean) values for $c.v_i = [0.01, 0.5]$
[e] In case of 2 available routes, the additional benefits (route choice) for sporadic drivers would be similar to those for commuters in same conditions

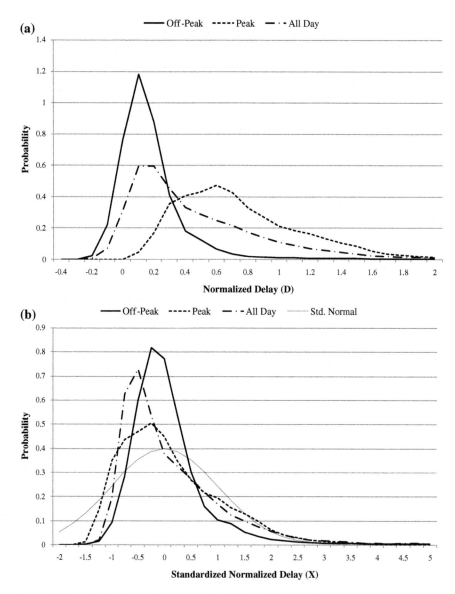

Fig. 7.4 Normalized delay distribution on the AP-7 turnpike accessing Barcelona, Spain. **a** Actual distributions. **b** Standardized distributions

7.5.1 Model Results

Both the mean and median values of the relevant results, obtained from the 10,000 replications of the model in each scenario, are presented in Table 7.5. Note that the results are, in fact, distributions of the possible information values, because in each

7.5 Numerical Examples

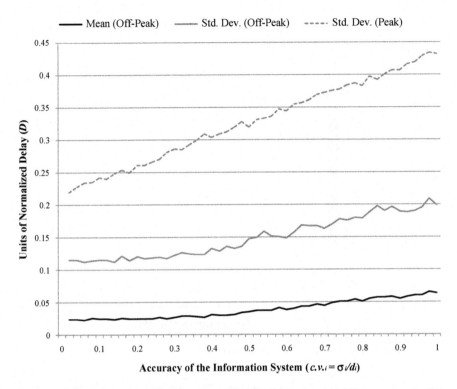

Fig. 7.5 Modified parameters for normalized delay distributions in case of two routes pooling service (i.e. massive information dissemination) *Note* Mean normalized delay for peak periods is invariant ($\mu_d = 0.72$)

trip, different circumstances occur that result in a different information value. Given the asymmetric shapes of these distributions (see Fig. 7.6) only providing the average value (as usual) could be misleading.

Table 7.5 shows that mean values are larger than median ones. This indicates a positive skewness of the distribution. The explanation for this is that, on many days, the performance of the freeway is as expected by the driver. In these "recurrent" conditions, little can be gained by changing departure times or route (because change of departure time is limited and routes are approximately in equilibrium). However, there are a smaller number of days where conditions differ significantly from those expected. In this "non-recurrent" condition, travel times in one route are abnormally long and the benefits of switching routes can be huge. The result is a distribution of the value of information with large frequencies for small values and a long tail for large values (see Fig. 7.6).

The skewness of the value of information distributions adds complexity in understanding how the driver perceives the overall value of the system. On average, the benefits drivers obtain from the system are represented by the mean value. However, for most of the trips they obtain values around the median. It is possible

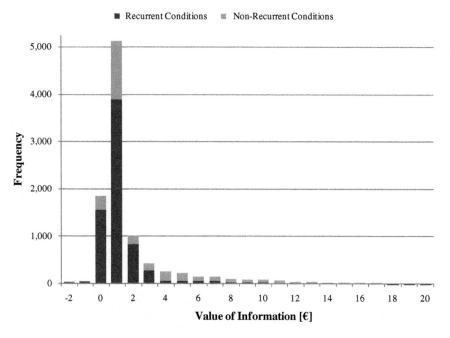

Fig. 7.6 Distribution of the value of information (Scenario 6)

that the more frequent values are the ones that drivers perceive more clearly. It could then be argued that the willingness to pay for the information (proxy for the driver perceived value) is more related to the median than to the mean. With these considerations, the results in Table 7.5 can be considered consistent with the willingness to pay results obtained through surveys (Khattak et al. 2003).

It is also clear from the results that in case of only one available route, the value of information is a fraction of the original scheduling costs. For the information having a significant value, significant original unreliability costs are necessary. This is the reason why the value of information for evening commute trips is very small.

In case of two possible routes, and limited information dissemination, information allows a better route choice, avoiding non-recurrent conditions on any route. This implies a reduction in both the effective travel time and the scheduling costs, and therefore the value of information is higher. This is not true when most of the drivers have information (i.e. massive dissemination). Then, the differential fact of having information does not imply a better route choice, because both routes will be approximately in equilibrium. However, the massive dissemination of information when two routes are available contributes in a reduction of the baseline unreliability, by reducing the travel time day-to-day variability. The value of this reduction is not explicitly stated in Table 7.5, but it can be easily obtained by comparing the original scheduling costs (i.e. without information) of Scenarios 6 and 7 (or 8 and 9, or 10 and 11). The reduced original scheduling cost of scenarios with massive dissemination of information is a result of the reduction of travel time variance due to most of the drivers having information. Then, even a particular driver without information would benefit from most of the others having it. These are the benefits of massive information systems, not perceived by any driver in particular but going to all of them.

The costs of stress are small and only significant when there is an important scheduled activity at the destination. In these scenarios, information reduces the stress costs by more than half. In contrast, on those trips without a rigid arrival time, travel time information adds a time limit (an expectation) which, when not entirely fulfilled, results in stress costs. That is the reason why the value of stress reduction is negative in these scenarios.

Scenarios 12 and 13, are aimed to represent average trip conditions on a metropolitan congested freeway. Considering that in these average conditions the lateness/earliness penalties are those reported on average by Small (1982) (see Table 7.3), the corresponding "P_L" value, for being consistent with expected utility theory, must be approximately 0.15. This could be represented by 80 % of commuters with a fixed arrival time, 14 % of commuters with flexibility and 6 % of sporadic drivers. As can be seen in Table 7.5, travel time information has a significant value under these average conditions.

Finally, note that small variations in the accuracy level result in equally small changes in the value of information. Given information, drivers adapt their decisions while maintaining their lateness probability. Then, small changes in accuracy (affecting travel time uncertainty in a few minutes) imply an equally small reduction of the buffer times, but not significant changes in the lateness/on-time probabilities, which would result in drastic variations for the information value. An exception to the previous statement could be the 2 route scenarios. In this case the worsening of the system accuracy may also imply a bad route selection. However, this "bad" decision only implies important costs when non-recurrent conditions prevail (i.e. abnormal delays in one route and significant travel time differences between routes). Therefore, costly bad decisions due to information can only happen when the system accuracy is very bad (i.e. errors of the same order of magnitude as travel times). To conclude, it can be stated that small accuracy improvements of information systems do not deserve being target objectives, but a minimum accuracy level (e.g. "$c.v._i$" < 0.2) is absolutely necessary.

7.6 Summary and Conclusions

Travel time uncertainty on highway trips imply important scheduling costs and stress for some drivers (ranging from 5 % up to 85 % of the time costs of the trip). The magnitude of these costs depends on the importance of reaching the destination on time, on the day-to-day travel time variance of the highway trip and on the previous knowledge the driver has of the travel conditions. This reveals two approaches to tackle travel time unreliability costs: (a) reducing the travel time variance, and (b) improving the driver knowledge through information. The former should be addressed by avoiding capacity fluctuations, mainly through traffic management strategies aiming to reduce the effects of incidents, in addition to an adequate management of highway demand. This would reduce both travel times in incident conditions and every day unreliability costs.

Information is the other option. The chapter proposes a method to quantify the value of highway travel time information systems, resulting from the reduction of unreliability costs. The method consists of two levels. First, a departure time selection and route choice model, based on a simplification of the expected utility theory, where it is only assumed that the driver has an accepted lateness probability. The model is simple, allowing a gain of knowledge about the insights and existing trade-offs, and it is flexible enough for being able to characterize most of the common travel scenarios. Second, an unreliability cost model accounting for scheduling costs and stress. The model takes into account the difference between planning and operative decisions the driver faces. This, for instance, implies limitations in the modification of departure time as a consequence of real-time information, generally not considered in the related literature.

The model assumes fixed wished arrival or departure times and no induced demand as a result of improved reliability. Therefore, its applicability is limited to these situations. Sporadic trips where the arrival/departure times are not fixed and the objective is more related to avoid congestion are out of the methodology scope.

Numerical examples for typical scenarios are presented, accounting for different types of trips (morning commute, evening commute, and sporadic trips), different traffic conditions (peak or off-peak periods), one or two available routes and massive or limited dissemination of information. Results indicate that when only one route is available, the value of information is only relevant when there is an important scheduled activity at the destination. For instance, a sporadic driver with an important meeting at a destination would obtain an average value of 4.78€ per trip from a travel time information system that provides historical (pre-trip) and real-time (on-trip) information. For the morning commuter, this value would be 0.74€. In case two equivalent routes are available, an additional value of 1€ is obtained from the system, as a result of a supported route choice decision.

The results also suggest that the willingness to pay for this information would be much smaller than its value. The extreme case would be the two route scenario with massive dissemination. The average driver would not perceive any significant value from information in this case (i.e. only 0.03€/trip). However, all drivers (even those uninformed) obtain a 0.99€/trip reduction of their scheduling costs as an indirect result of information. An additional reduction of 0.56€/trip is obtained if the driver is explicitly informed. This means that the average value of more than 1.5€/trip is not perceived, at all, by the driver.

Finally, it is also confirmed that accuracy of the information system is not critical for obtaining these values. Accuracy improvements, given a minimum accuracy level (max. relative error below 20 %), result in small additional benefits.

Appendix 7A: Variance of Service Time in Centralized Versus Decentralized Oversaturated Queuing Systems

In systems where demand, "λ", exceeds capacity "μ", the variance of service time is proportional to the variance of the demand/capacity ratio, "$y = \lambda/\mu$". Consider two independent (i.e. decentralized) systems. The variance of the total service time would be proportional to the variance of "$y_1 + y_2$".

$$Var(y_1 + y_2) = \xrightarrow{independence} = Var(y_1) + Var(y_2) = \xrightarrow{y_1 \approx y_2 \approx y} = 2 \cdot Var(y) \quad (7.19)$$

In case of centralized systems, the previous variance would be proportional to the variance of "$\frac{\lambda_1 + \lambda_2}{\mu_1 + \mu_2} \approx y$". Therefore centralization reduces the service time variance to half.

References

AASHTO. (1977). *A manual on user benefit analysis of highway and bus transit improvements.* Washington, D.C.: American Association of State Highway and Transportation Officials.
Asensio, J., & Matas, A. (2008). Commuters' valuation of travel time variability. *Transportation Research Part E, 44*(6), 1074–1085.
Avineri, E., & Prashker, J. N. (2003). Sensitivity to uncertainty: The need for a paradigm shift. In *Proceedings of the 82nd Transportation Research Board Annual Meeting*, Washington, D.C.
Bates, J., Polak, J., Pones, P., & Cook, A. (2001). The valuation of reliability for personal travel. *Transportation Research E, 37*(2–3), 191–229.
Ben-Akiva, M., De Palma, A., & Isam, K. (1991). Dynamic network models and driver information systems. *Transportation Research A, 25*(5), 251–266.
Ben-Elia, E., & Shiftan, Y. (2010). Which road do I take? A learning-based model of route-choice behavior with real-time information. *Transportation Research A, 44*(4), 249–264.
Bonsall, P. (2003). Traveller behaviour: Decision making in an unpredictable world. *Journal of Intelligent Transportation Systems, 8*, 45–60.
Chorus, C. G., Arentze, T. A., Molin, E. J. E., Timmermans, H. J. P., & Van Wee, B. (2006). The value of travel information: Decision strategy-specific conceptualizations and numerical examples. *Transportation Research B, 40*(6), 504–519.
Chrobok, R., Kaumann, O., Wahle, J., & Schreckenberg, M. (2004). Different methods of traffic forecast based on real data. *European Journal of Operational Research, 155*, 558–568.
De Neufville, R. (1990). *Applied systems analysis: Engineering planning and technology management.* McGraw-Hill.
Emmerink, R. H. M., Nijkamp, P., & Rietveld, P. (1995). Perception and uncertainty in stochastic network equilibrium models: An alternative approach. TI 95-159, Tinbergen Institute, Amsterdam-Rotterdam, The Netherlands.
Ettema, D., & Timmermans, H. (2006). Costs of travel time uncertainty and benefits of travel time information: Conceptual model and numerical examples. *Transportation Research C, 14*(5), 335–350.
Fosgerau, M., & Karlström, A. (2010). The value of reliability. *Transportation Research B, 44*(1), 38–49.
Jodar, V. (2011). Valor de la informació del temps de viatge per carretera. Master thesis directed by F. Soriguera (in Catalan). Barcelona Civil Engineering School, Barcelona-Tech.

Khattak, A., Polydoropoulou, A., & Ben-Akiva, M. (1996). Modeling revealed and stated pretrip travel response to advanced traveller information systems. *Transportation Research Record, 1537*, 46–54.

Khattak, A. J., Yim, Y., & Prokopy, L. S. (2003). Willingness to pay for travel information. *Transportation Research C, 11*(2), 137–159.

Lam, T. C., & Small, K. A. (2001). The value of time and reliability: Measurement from a value pricing experiment. *Transportation Research E, 37*(2–3), 231–251.

Levinson, D. (2003). The value of advanced traveler information systems for route choice. *Transportation Research C, 11*(1), 75–87.

Li, R., Rose, G., & Sarvi, M. (2006). Using automatic vehicle identification data to gain insight into travel time variability and its causes. *Transportation Research Record, 1945*, 24–32.

Mahmassani, H. S., & Chang, G. L. (1986). Experiments with departure time choice dynamics of urban commuters. *Transportation Research B, 20*(4), 297–320.

Mahmassani, H. S., & Chen, P. S. T. (1991). Comparative assessment of origin-based and en route real-time information under alternative user behaviour rules. *Transportation Research Record, 1306*, 69–81.

Mahmassani, H., & Jou, R. C. (2000). Transferring insights into commuter behavior dynamics from laboratory experiments to field surveys. *Transportation Research A, 34*(4), 243–260.

Nadarajah, S., & Kotz, S. (2008). Exact distribution of the max/min of two Gaussian random variables. *IEEE Transactions on Very Large Scale Integration (VLSI) Systems, 16*(2), 210–212.

Noland, R. B., & Polak, J. W. (2002). Travel time variability: A review of theoretical and empirical issues. *Transport Reviews, 22*, 39–54.

Noland, R. B., & Small, K. A. (1995). Travel-time uncertainty, departure time choice, and the cost of morning commutes. *Transportation Research Record, 1493*, 50–158.

Palen, J. (1997). The need for surveillance in intelligent transportation systems. *Intellimotion* (Vol. 6:1, pp. 1–3, 10). University of California PATH, Berkeley, CA.

Small, K. A. (1982). The scheduling of consumer activities: Work trips. *The American Economic Review, 72*, 467–479.

Torné, J. M., & Soriguera, F. (2012). Active traffic management strategies: State of the art review. In *Proceedings of the X Congreso de Ingeniería del Transporte, Granada*. (in Spanish).

Train, K. (2003). *Discrete choice methods with simulation*. Cambridge University Press.

Turner, S. M., Eisele, W. L. Benz, R. J & Holdener, D. J. (1998). *Travel time data collection handbook*. Research Report FHWA-PL-98-035. Texas Transportation Institute, Texas A&M University System, College Station, Tx.

Van Lint, J. W. C., van Zuylen, H. J., & Tu, H. (2008). Travel time unreliability on freeways: Why measures based on variance tell only half the story. *Transportation Research A, 42*(5), 258–277.

Vickrey, W. S. (1969). Congestion theory and transport investment. *American Economic Review, 59*(2), 251–261.

Walker, J., & Ben-Akiva, M. (1996). Consumer response to traveler information systems: Laboratory simulation of information searches using multimedia technology. *Intelligent Transportation Systems Journal, 3*(1), 1–20.

Wardrop, J. G. (1952). Some theoretical aspects of road traffic research. *Proceedings of the Institute of Civil Engineers, 2*, 325–378.

Printed by Printforce, the Netherlands